COLONIAL SEEDS IN AFRICAN SOIL

The Environment in History: International Perspectives

Series Editors: Dolly Jørgensen, *University of Stavanger*; Kieko Matteson, *University of Hawai'i at Mānoa*; Christof Mauch, *LMU Munich*; Helmuth Trischler, *Deutsches Museum, Munich*

eseh european society for environmental history

Rachel Carson Center
ENVIRONMENT AND SOCIETY

VOLUME 18
Colonial Seeds in African Soil: A Critical History of Forest Conservation in Sierra Leone
Paul Munro

VOLUME 17
Hazardous Chemicals: Agents of Risk and Change, 1800–2000
Edited by Ernst Homburg and Elisabeth Vaupel

VOLUME 16
Planning for the Planet: Environmental Expertise and the International Union for Conservation of Nature and Natural Resources, 1960–1980
Simone Schleper

VOLUME 15
Changes in the Air: Hurricanes in New Orleans from 1718 to the Present
Eleonora Rohland

VOLUME 14
Ice and Snow in the Cold War: Histories of Extreme Climatic Environments
Edited by Julia Herzberg, Christian Kehrt and Franziska Torma

VOLUME 13
A Living Past: Environmental Histories of Modern Latin America
Edited by John Soluri, Claudia Leal and José Augusto Pádua

VOLUME 12
Managing Northern Europe's Forests: Histories from the Age of Improvements to the Age of Ecology
Edited by K. Jan Oosthoek and Richard Hölzl

VOLUME 11
International Organizations and Environmental Protection: Conservation and Globalization in the Twentieth Century
Edited by Wolfram Kaiser and Jan-Henrik Meyer

VOLUME 10
In the Name of the Great Work: Stalin's Plan for the Transformation of Nature and Its Impact in Eastern Europe
Edited by Doubravka Olšáková

VOLUME 9
The Nature of German Imperialism: Conservation and the Politics of Wildlife in Colonial East Africa
Bernhard Gissibl

For a full volume listing, please see the series page on our website: http://berghahnbooks.com/series/environment-in-history.

Colonial Seeds in African Soil

A Critical History of Forest Conservation in Sierra Leone

Paul Munro

berghahn
NEW YORK • OXFORD
www.berghahnbooks.com

First published in 2020 by
Berghahn Books
www.berghahnbooks.com

Library of Congress Cataloging-in-Publication Data

Names: Munro, Paul, 1958- author.
Title: Colonial seeds in African soil : a critical history of forest conservation in Sierra Leone / Paul Munro.
Other titles: Environment in history ; b. 18.
Description: New York : Berghahn Books, 2020. | Series: Environment in history: international perspectives ; volume 18 | Includes bibliographical references and index.
Identifiers: LCCN 2019043784 (print) | LCCN 2019043785 (ebook) | ISBN 9781789206258 (hardback) | ISBN 9781789206265 (ebook)
Subjects: LCSH: Forest conservation--Sierra Leone. | Forest conservation--Political aspects--Sierra Leone. | Sierra Leone--Colonial influence.
Classification: LCC SD414.S5 M86 2020 (print) | LCC SD414.S5 (ebook) | DDC 333.751609664--dc23
LC record available at https://lccn.loc.gov/2019043784
LC ebook record available at https://lccn.loc.gov/2019043785

British Library Cataloguing in Publication Data

A catalogue record for this book is available from the British Library

ISBN 978-1-78920-625-8 hardback
ISBN 978-1-78920-626-5 ebook

Contents

Illustrations

Acknowledgements

First and foremost, a thank you is owed to the Australian Government with its Australian Postgraduate Award (APA) scholarship, without which I would not have been able to have completed the research for this book. Hopefully, the Government will continue to invest in tertiary education and academic research in future. Also, a thank you is owed to the School of Geography at the University of Melbourne, who housed me during my research, provided me with resources and, most importantly, provided an academic environment where critical discussions could occur. A special thanks is also merited for my fellow travellers during my doctoral research, who helped to keep me sane during the long hours of writing up my thesis, even if it did often result in extended periods of procrastination (you know who you are). This particularly includes Dr Greg van der Horst, my partner in academic crime (and frequent co-author), when it came to trying to understand the complex dynamics of forests in Sierra Leone.

I am of course greatly indebted to Professor Simon Batterbury (University of Melbourne) and Professor Tony Binns (University of Otago) for their constant support. Tony's long history of academic research in Sierra Leone (40 years and counting!) provided me with invaluable insights into the past I was trying to reach. Simon was an inspiring supervisor, who went against much of the grain and prioritised generous and mindful supervision as a core praxis of his university work. In particular, he fostered a spirit of collaboration and support among his PhD student cohort. I remember just after starting my PhD I was crammed into his battered Toyota Prius with a bunch of other PhD students for a road trip to a political ecology workshop at the University of Wollongong. Conversations that started on this trip, with Simon and other PhD students, continued throughout my studies and formed the background of intellectual discussion and engagement that fed into this book. His excellent supervision was (finally!) formally recognised in 2019, with a graduate student supervision award from the University of Melbourne.

I am also greatly indebted to the University of New South Wales (UNSW), my current employer. They took a chance with me; first taking me on as a lecturer early in my career, then providing me with their Scientia Research Fellowship, which, among many things, provided me with the research time and resources necessary to complete this book. I am extremely fortunate to be part of the Environment and Society Group within the School of Humanities and Languages at UNSW. My colleagues and friends are both academically brilliant and great fun to be around. We have found a great balance in combining rigorous academic debate with lots of jokes and laughter. They have inspired and contributed to this book in many ways.

The hard-working staff at libraries and archives around the globe were amazing in pointing a disorientated researcher in the right direction so that he could delve into Sierra Leone's forest past. All of the research participants were amazing, allowing me to visit their personal histories as well as supplying me with invaluable knowledge and materials. Also, a special thanks to Associate Professor John Dargavel (Australian National University) for sharing with me his notes on Charles Lane-Poole, which were critical for many parts of the book. My organisation in Sierra Leone, Energy For Opportunity, also played a critical role in supporting my research and providing me with some good times. Thanks Simon Willans! This also includes the great work of Mohamed, Aiah, Georjet, Isaiah, Albert and Marah – all excellent field researchers in helping me to realise different parts of this project.

A big thanks to Professor Dolly Jorgensen (University of Stavanger), my editor, who has provided fantastic support throughout. She gave great feedback, showed great patience (even when I missed deadlines!) and found excellent reviewers (anonymous and Professor Gregory Maddox), who provided constructive feedback on earlier versions of the book's manuscript. The final book is so much stronger thanks to the time Dolly and the reviewers have spent in going through my work.

I am also indebted to the European Society for Environmental History (ESEH) and the Rachel Carson Center (RCC) for their role in supporting this book series, 'Environment in History'. It is a privilege for my book to sit among such a phenomenal collection of environmental history scholarship; this, of course, includes a big thanks to Berghahn Books for publishing the series and supporting my book and research. A special thanks to Mykelin Higham and Caroline Kuhtz at Berghahn Books for their fantastic work in helping to finalise the manuscript.

A very special acknowledgement goes to my extended family in Mexico, who has provided me with constant and unwavering support and understanding: Maria del Carmen, Pancho, Anita, Maca, Mauricio and Takeshi. To my own direct family – Jim, Marilyn, Rob (aka Raúl) and Kylie – thanks for your steadfast support. You have provided me with a constant source of inspiration and assistance throughout eclectic studies, international (mis)adventures and stumbles through different 'career' paths.

There have been two people that have been 'forced' to live with the research underlying this book on a daily basis. This has included travelling with me across four different continents and moving houses more times than I care to remember. Marilu, words cannot express my thanks for your unwavering support and belief that I could complete this book despite the many obstacles *we* faced. The terms 'hard' and 'challenging' do not do justice to the collective efforts of both of us completing PhDs and carving out careers in academia. On the 10 May 2010, three months after starting my PhD journey, Makena Munro Melo – my beautiful daughter – was born. If a household trying to complete two PhDs wasn't hectic enough, Makena was determined to make her own contribution: for her first 12 months she decided that sleeping was just something that other babies did. In all seriousness, though, Makena provided me with the inspiration to complete this book. She made me laugh and taught me a whole new way of looking at the world. Marilu and Makena, I dedicate this book to both of you.

Introduction

5 June, 2010: 'Trees for Life'

The 5th of June is a prominent day on Sierra Leone's calendar – it is National Tree Planting Day. On this day in 2010, Sierra Leone's Vice President, Samuel Sam-Sumana, is trekking up the slopes of the Western Area Peninsula Mountains on the outskirts of Freetown, the country's capital city. He is accompanied by the Minister for Agriculture, Forestry and Food Security (MAFFS), the Minister for Land, Country Planning and the Environment (MLCPE), the Minister for Energy and Water Resources (MEWR) and a senior representative from the Environmental Protection Agency (EPA). In addition to this governmental collective, representatives from the European Union, staff from large environmental non-government organisations (NGOs) and journalists from Sierra Leone's major newspapers are in tow.

This is not the Vice President's first hike up the Western Area Peninsula Mountains. In 2008, soon after his party was elected into government, he trekked up for his first National Tree Planting Day as Vice President,[1] and since this time the Vice Presidential pilgrimage up the mountain slope has become almost an annual tradition. The choice of the Peninsula Mountains as the site for tree planting is iconic, as the visibility of the forested mountains from the capital city Freetown provides symbolic and political capital, emphasising his conservation credentials. The Vice President's 2010 trek is in support of a five year project to improve the management of the Forest Reserve that traverses the forested mountains. The project, funded by the European Union, is being implemented by the German NGO Welthungerhilfe and was initiated in 2009.

The Vice President and his following entourage finally reach their destination: the physical boundary of the Forest Reserve, where a formal ceremony is initiated. The ceremony begins with a speech by Vice President Sam-Sumana, starting with the leitmotif that over 60 per cent of Sierra Leone's forests have been depleted and destroyed by human activities and that it is incumbent on 'the government of Sierra Leone to protect the forest' and prevent it from being 'destroyed by bad people'.[2] The peroration of the speech aligns with the day's theme: 'the government is championing the maximization of trees for beneficial purposes' to ensure sound environmental management in Sierra Leone.[3] After this initial monologue, the Ministers take their turn to extend the Vice President's jeremiad in their speeches, noting that deforestation is not only caused by people who cut down trees but also by those who purchase the timber.[4]

The ceremony, however, is not just an exercise in oration. The Vice President, after his speech, moves to lay the foundation for the first pillar for the re-demarcation of the Forest Reserve, demonstrating the government's commitment to the environment and to Welthungerhilfe's work. More political support will come in future, with the Sierra Leonean Government re-gazetting the Forest Reserve as a National Park in June 2013, the aim being to provide greater protection for wildlife in the area by further restricting unauthorised human access. After the foundation pillar is laid, the crux of the ceremony is finally reached, with the Vice President planting a *Gmelina arborea* tree – a symbolic gesture to mark not only National Tree Planting Day but also the start of tree planting season. On this day, and over the next three months, thousands of *Gmelina arborea* (known locally as *Yemane*) and other, mainly non-native, fast growing trees will be planted by the Sierra Leonean Government, environmental NGOs, development agencies and school children, among others, as a nationwide attempt to 'reclothe' the Sierra Leonean landscape with trees. The theme for the 2010 National Tree Planting day is 'Trees for Life'.

A Political Ecology of Sierra Leone's Forest Conservation History?

The introductory narrative is a concise exemplar of the political ecology of forest conservation in Sierra Leone, the topic of this book. At first glance, the vignette appears to be focused on Sierra Leone's forest future, or on its forest present, with the imperative of improving the country's forest conservation through planting trees and re-establishing protected areas. However, one does not need to dig too deep into the narrative to uncover historical references. The first question to perhaps ask is why is Sierra Leone celebrating National Tree Planting Day on the 5th of June every year? Almost everywhere else in the world that same day is designated as being World Environment Day. Yet in Sierra Leone, the day was renamed to its current moniker in 1985, and afforestation activities have ensued, with one Sierra Leonean forester (perhaps ambitiously) estimating that between 1985 and 2010 the country's Forest Department had succeeded in disseminating over 40 million tree seedlings around the country as a result of annual intense tree planting season initiatives.[5] Tree planting appears to have become synonymous with environmental conservation in Sierra Leone, but what does this mean? Why is this the case?

The Vice President's choice of tree, *Gmelina arborea*, is in itself steeped in history and ambition. This book traces the genealogy of the sapling that the Vice President planted in 2010 back to a packet of seeds shipped to the Sierra Leonean colonial Forestry Department from India in 1920. Conservation day, in Sierra Leone, is therefore marked by the government advocating, almost without knowing, the introduction of foreign tree species. It is an act that is not limited to one day. Non-native tree species, and *Gmelina arborea* in particular, seem to be favoured for tree planting programmes due to their fast growth rates and rapid germination. This, therefore, proffers a series of questions including: does the Vice President know where his *Gmelina arborea* sapling originally came from? How does he know that 60 per cent of the forests in Sierra Leone have been destroyed? But, most importantly, what sort of forested landscape does the Sierra Leonean Government (and its predecessor the British Colonial Government) desire to create and why?

The rhetoric of the day casts Sierra Leone's environment as being heavily degraded, backed by speculative statistics about Sierra Leone's forest past, and argues the need for demarcated Forest Reserve boundaries as a tool to arrest deforestation. This rhetoric is all the product of Sierra Leone's forest history. Even the mountainous forested slopes, where the tree planting ceremony was held, has long been a focal point for environmental conservation discourse and activity in Sierra Leone. Since the 1790s, concerns about its deforestation have been a key source of discussion among senior colonial officials,[6] culminating in the establishment of a forest reserve in 1916. The founder of the reserve was Sierra Leone's first Conservator of Forests (i.e. Forestry Department Head) Charles Lane-Poole. Like Vice President Sam-Sumana, Lane-Poole described the establishment of Forest Reserves as being an urgent necessity for Sierra Leone as '99% of the colony's forests had been destroyed by the activities of the native population'.[7] The 2010 National Tree Planting Day is not the result of an apolitical desire to enhance Sierra Leone's forest cover, but rather it is the product of the 'forest conservation' environmental history shaping the discourses and praxis surrounding the country's forests.

This book unravels the complex forest conservation history in Sierra Leone in order to provide a better understanding of how contemporary forest conservation has emerged. As Bruce Braun has noted, there is an ultimate need to recognise forests as being inherently epistemic, cultural and political spaces.[8] National Tree Planting Day in Sierra Leone is not simply the product of an innate desire to have more trees on the Sierra Leonean landscape; it is the product of a complex history of ideas and practices. Likewise, forest conservation should not be viewed simply as a system of managing and protecting a collection of trees, rather it needs to be viewed as a social process shaped by the complex political economy. A *critical* history of forest conservation, therefore, is not simply about recounting policies and initiatives to manage Sierra Leone's forests; it is about attending to the complicated social, economic and political contexts that have shaped approaches to forest management initiatives over time.

In this book, I adopt an interpretative analytical approach to understanding Sierra Leone's forest history. It is grounded in a critical realist epistemology in the sense that it recognises that although the forests of Sierra Leone represent a physical entity and are the site of socio-human interactions, their 'conservation' is ultimately realised through a mixture of socially constructed perceptions and institutional forces. In doing so, the book draws upon insights from the broad fields of 'environmental history' and 'political ecology'. Environmental history, as a field, has helped to reorientate historical narratives to ensure that the progression of human–nature interactions becomes a key focal area,[9] thus providing a distinctive way of looking at the past to provide a context for environmental change.[10] Works in political ecology have had more of a focus on understanding how seemingly localised environmental situations are linked to the broader political economy, both materially and discursively.[11] As such, it is a conceptual approach that seeks to unravel the forces at work in shaping environmental access, management and transformation.[12] Political ecology, therefore, is not so much focused on reframing history (as environmental history is) but rather on interrogating key themes and apparent contradictions in Sierra Leone's forest history to reveal the complexities behind forestry conservation programmes and policies, including power, politics and multi-scalar forces.

The approach of this book sits within the broader collection of political ecology works that Paul Robbins categorises as engaging with the 'Conservation and Control Thesis',[13] the focus being on revealing how environmental conservation knowledge and practice in many parts of the world has less to do with science and an empirically grounded understanding of environmental change and more to do with the competing claims emerging from political and bureaucratic institutional arrangements. Much of the focus of political ecology, in this area, therefore, has been on understanding the historical processes of colonialism and its influence in shaping contemporary structural relations between the state, civil society and markets.[14] It is important to note that the conservation and control thesis is not against the defence (and conservation) of ecological systems, biodiversity and wildlife. Rather, it emphasises the degree to which such objectives have historically failed, primarily because approaches to conservation have systemically disenfranchised rural communities and have enforced the desires and benefits of elites who only have a limited understanding of ecosystem process, landscapes or local places.[15] There is thus a normative purpose to political ecology in this area in that its objective is to radically reorientate conservationist thinking.

A key concern of political ecology scholarship has been to challenge hegemonic ideas about the environment, and therefore many works have used an historical lens to unpack discursive legacies of conservation. There is thus an entangled political and environmental history approach – what has been termed Historical Political Ecology – as Karl Offen usefully summarises: 'historical political ecology can be characterized as a field-informed interpretation of society-nature relations in the past (e.g. material, ideological, legal, spiritual), how and why those relations have changed (or not changed) over time and space, and the significance of those interpretations for improving social justice and nature conservation today.'[16] Thus, in many ways it represents a 'history of the present',[17] as this approach has a focus on conceptual genealogies to formulate an understanding about how current geographical, social and political relations and conditions have emerged.[18] Tim Forsyth notes that such an approach involves interrogating narratives 'using poststructuralist and historical analysis to indicate how environmental explanations carry many hidden normative values'.[19] By understanding the philosophical and social roots of a hegemonic discourse and/or material practices we are in a better position to deconstruct them. This has been a particularly popular approach in political ecology works seeking to critically challenge forestry and forest conservation activities in postcolonial states.[20] It is the approach that this book has adopted for its analysis; it aims to destabilise current approaches and ideas surrounding forest conservation in Sierra Leone by illustrating their close relationship to earlier colonial constructions of the African environment rather than by exploring empirical grounded knowledge of Sierra Leone's forested environs – constructions that led to the establishment of particular institutional arrangements to promote forest conservation.

Given this approach, in this book I mobilise the idea of 'environmental history' (or perhaps more accurately 'environmental histories') in two distinct ways to historically understand forests and forestry in Sierra Leone during the colonial era. The first is to build on the broader academic project of environmental history. While, as Douglas Wiener poignantly reminds us,[21] environmental history is a relatively eclectic field, it nevertheless is at its greatest strength when it helps to elevate environmental questions into our historical narrative and subsequently into public debate.[22] As Tom Griffiths succinctly opines, in history writing the environment problematically has often just been a 'stage and setting for the human drama'. To correct this, environmental historians therefore seek to enrol 'an active nature in the narrative'.[23] Environmental history thus offers a different lens through which to understand the past, providing new perspectives for engaging in historical and contemporary debates.[24] This is a particularly important project in Africa, as the environmental history field has thus far arguably been biased towards research from Europe and North America, due to the historical threads of its intellectual foundations. The environmental history project in Africa is ultimately much more nascent and threadbare, [25] especially in West Africa,[26] and therefore this book is part of an effort to move the field into new grounds. Indeed, despite the vast majority of Sierra Leone's population relying direct on its forests for either livelihoods or commercial opportunities,[27] as well as the presence of an active Forestry Department for more than 100 years, there has been no book dedicated to its forest history.

Second, it is also important to appreciate that, beyond academia, environmental histories already exist in many forms in Africa. Existing knowledges and stories of environments and environmental change already are constructed across different scales and have had implications in terms of policies and project interventions. At the national and transnational scale, these environmental history narratives have often been problematic. As Diane Davis has noted in her research in North Africa, colonial powers were often involved in constructing certain kinds of environmental histories to justify significant parts of their colonial projects. And, despite the 'colonial science' rhetoric, often these constructed histories were far from being 'scientifically' accurate. Even more problematic, these colonial environmental history narratives, rather than being questioned or investigated by the postcolonial states, have often become the dominant postcolonial environmental history as well. [28] James Fairhead and Melissa Leach come to similar conclusions with their research on West Africa's forests, describing these as 'false forest histories'[29] – stories of environmental change that have emerged from populist hegemonic discourse, not from actually 'reading' the landscape carefully.[30] Dominant environmental historical narratives that have emerged in the colonial era in Africa have thus been sullied by confusions, misunderstandings, misinterpretations and falsehoods.[31] James McCann simply describes these as 'apocryphal environmental narratives' that have emerged across Africa telling a 'plausible story' of environmental degradation that places (quite problematically) blame on poor, rural populations.[32] Colonial environmental histories, ultimately, are still dominant narratives in many parts of Africa.

As such, the work of this book, and arguably the broader project of political ecology- informed environmental history, is about attending to the competing and conflicting environmental histories that already exist, whether they are presented in the form of academic works, formal government policy documents or popular discourse. As William Beinart notes, 'new' environmental histories of Africa effectively adopt an 'essentially corrective and anti-colonial approach' that tends to emphasise 'African initiative in the face of European conquest and capitalist exploitation'.[33] Environmental history in Africa is thus very much a radical project. It is not just about telling the environmental history of Sierra Leone's forests but also understanding the genealogy behind why different environmental histories already exist. And why they might be problematic.

Colonial forestry history in Africa, and other parts of the world, has been an increasingly active area of scholarship in recent years, garnering the attention of a range of historians, geographers, anthropologist and foresters. In terms of grandiose arguments about the legacy of colonial forestry, there are two key broad threads that sit somewhat in tension. The first is a notion of colonial foresters being early environmentalists, laying down the institutions, laws and philosophies that would form a critical basis for the environmental movement that would emerge in the latter part of the twentieth century. Foresters, among other things, were the progenitors of sustainable resources, and their philosophies and dedication to protected area developments would later become mainstream.[34] The second thread paints colonial foresters as being a much more destructive force, arguing that after arriving (from Europe) to the tropics, they misunderstood and misread local landscapes, dismissed existing knowledge systems as ignorant and ultimately created policies that had negative (and sometimes devastating) impacts on local populations and environments.[35]

These two threads, at first glance, seem quite disparate – foresters as pioneers of environmentalism, or as accomplices of imperial devastation. However, in this book I will look to bring the two threads together, specifically with the question: *what* kind of environmentalism did colonial forestry help to craft in Africa? An important consideration for this is a recognition of how the notion of what is 'environmentalism' is contested. Joan Martinez-Alier's typologies of environmentalism are instructive in this regard: environmentalism can be characterised as the protection of economic resources (e.g. timber in forests); it can be characterised as the protecting of intrinsic values (e.g. forests as a place for other species); and it can be characterised as an environmental justice issue (e.g. poorer groups' rights to have control over their local forest areas).[36] In this book, I argue that Sierra Leonean forest conservation can best be understood as being the tension of such environmentalisms. Colonial forestry undoubtedly helped to bring a utilitarian (colonial) environmental ethic to Africa, as well as arguably providing a foundation for later wilderness conservation initiatives such as the creation of national parks. However, this imposed environmentalism has created numerous environmental injustice outcomes across the continent.

In bringing a critical eye to the impacts of colonial forestry in Sierra Leone, it is important to take care in not overstating the influence of colonial forestry. As Brett Bennett has noted, many historians have often ascribed a hegemonic-like power to foresters, failing to 'take into account the political, social, economic, and environmental constraints limiting the power of professional foresters'.[37] Thus we should not 'conflate policy intent with practical outcome':[38] foresters had grand ideas about how the African landscape should be managed and utilised; however, their influence was curtailed in many ways. British Imperialism was not a totalising force, and colonial forestry even less so. While colonial foresters were able to trumpet their self-claimed superior knowledge of forestry and attempted to justify their activities in a broader scientific, economic and moral framework, their ability to realise their vision in sub-Saharan Africa was hindered, as European control over the African landscape was never absolute.[39] While the European powers had been militarily successful in securing territories in Africa at the end of the nineteenth century, control and access to resources involved a much longer process of negotiation with local Indigenous elites. This was especially the case in West Africa, where, unlike other parts of Africa and the colonial world where a settler approach was favoured, an indirect rule and local autonomy system was adopted by the British and French Governments.[40] Thus, instead of directly claiming land as European property in this region, they instead predominantly relied on co-opting existing African chiefdom governance structures as a means of gathering taxes and overseeing land-tenure arrangements. This was especially the case in rural areas. This meant that colonial governments were in a weak position with respect to establishing 'environmental hegemony' across the region.[41] Sara Berry describes this as being 'hegemony on a shoestring' – the British Empire wanted to control its colonies; however, it wanted to invest as little as possible to realise this desire.[42] Thus, while the powers of chiefs were subordinated to those of colonial state authority – for example, in relation to duties of tax collection and labour recruitment for military and public works – they were often greatly increased in relation to their subjects. Therefore, the colonial refashioning of the chieftaincy infused executive, legislative and judicial powers of 'customary' authority over rural areas.[43] Chiefs in colonies such as Sierra Leone, Nigeria and Ghana were able to carve out an important and influential position in the colonial governance system, and, as will be shown, they were in turn able to frustrate processes like the Forestry Department's efforts to achieve 'scientific forestry'.

Further compounding the weakness of indirect rule was the fact that the short colonial experiment in Africa occurred during a globally volatile period – both economically and politically. Between the establishment of Forestry Departments across Africa during the early 1900s and independence being realised by most colonies during the 1950s and 1960s, the world experienced two major wars that were interceded by a period of economic downturn. Preoccupation with such events ultimately had reverberations on how much time and capital could be spent on securing the natural resources of the Empire's colonies.

Thus, as the book title metaphorically notes, the weakened *seeds* of colonial forestry were planted in African *soil*, and therefore the contemporary environmentalism *tree* that has emerged is shaped by this tension: European ideas and praxis being imposed in an African context. This tension, I argue, is ultimately what continues to characterise forest conservation in contemporary Sierra Leone and much of sub-Saharan Africa: postcolonial elites (governments, NGOs, aid donors etc.) are still very much trying to develop a colonial ideal of conservation in a defiant African context. When the Sierra Leonean Vice President Samuel Sam-Sumana planted his *Gmelina arborea* tree while declaring there is a need to protect the forest from an ignorant local population, he was continuing the quest laid out by the colonial Forestry Department some 100 years earlier – a Sisyphean quest to displace African practice with a European epitome. And, therefore, contemporary forest conservation in Sierra Leone and much of Africa ultimately embodies a range of conflicting pre-colonial, colonial and postcolonial ideals.

To understand these tensions that characterise contemporary forest conservation in Africa – as well as to stretch the book title's metaphor as far as possible – the following chapters will examine the *seeds* of colonial forest conservation in Sierra Leone.

Chapter 1 provides a contextual early history of Sierra Leone, detailing its early colonial settlement and the broader imperial processes that led to the establishment of the Sierra Leonean Forestry Department. It includes an overview of Sierra Leone's vegetation and wildlife geography. Chapter 2 focuses on the first major policy and programme area of the Forestry Department: reservationism, the haphazard process that colonial foresters underwent in trying to transform large sections of Sierra Leone's land and forest resources into formalised Forest Reserves – a process that was 'supposed' to secure a large forest estate for the Forestry Department. The chapter shows how reservationism as a process – but even more importantly as an imaginary – has arguably had a profound impact in terms of how conservation is conceived of and approached in Sierra Leone; most notably, the idea that forest in Sierra Leone can only be protected through a protected area management approach. Chapter 3 focuses on the Department's second major programme: plantations. If Forest Reserves were about securing forest resources, then plantations were about ensuring that forest-land was filled with 'useful' trees for colonial forestry activities, and therefore the chapter maps out the history and experiments of changing the forestscape of Sierra Leone by planting more trees en masse. This chapter, in particular, looks at the imported (from South Asia) tree species of *Gmelina arborea*, known locally as *Yemane*, and how it came to dominate the Sierra Leone forestscape due to a range of ecological and social dynamics. The fourth chapter, 'Exploitation', focuses on the third policy of Sierra Leone's Forestry Department – the one that completes the logic of colonial forestry. Reservation secured land and plantations helped to ensure it was full of 'useful' trees, while exploitation focused on harvesting these spaces for colonial profit. Chapter 4 thus explores the struggles of the Forestry Department to transform its operations into a profit-making venture, a key colonial imperative. The chapter explores the massive boom in forest exploitation activities by the Forestry Department that occurred during and after World War II, kick-starting a burgeoning commercial forestry industry in the country. This is followed by an analysis of its decline during the 1970s and 1980s due to the changing political economy of Sierra Leone.

Chapter 5, 'Wildlife Conservation', focuses on the Department's final programme, one that emerged in the postcolonial era although with earlier colonial threads. It is a programme that represents a distinct shift in philosophy for the Department in that forests were increasingly being promoted as places of intrinsic (rather than just utilitarian) value – as places where non-human species live. This chapter is a striking example of the dynamics of wildlife conservation in West Africa, where, unlike other parts of sub-Saharan Africa, wildlife conservation became a subprogramme of broader forestry governance, rather than acting as its own department. This chapter therefore looks at the tensions, politics and processes that were involved in overlaying wildlife conservation initiatives upon the institutional infrastructure of colonial forestry. Finally, the *epilogue* then reflects on what this history means for the contemporary endeavour of environmentalism and forest conservation in Sierra Leone.

The research for this book was conducted between 2009 and 2012, and its main source of information was derived from a wide range of library and archival sources from Sierra Leone (i.e. Fourah Bay College (University of Sierra Leone) Library, Freetown Public Library, Njala University College Library, Sierra Leone National Archives); the United Kingdom (i.e. Adeilad Deiniol Library, Bodleian Library, British Library, British Library of Political & Economic Science, Foyle Reading Room, Harrison Institute, Radcliffe Science Library, School of Oriental and African Studies (SOAS) Library, Senate House Library, Sherardian Library of Plants, The British Library, The National Archives, University College London (UCL) Library); and Australia (i.e. National Library of Australia, New South Wales State Library). Materials from this research included diaries, colonial forestry reports and colonial forestry communications, among other work. This archival research was complemented with eleven oral histories, including that of a former Chief Conservator of Forests, colonial foresters who worked in Sierra Leone in the 1960s and 1970s, primatologists who worked on wildlife conservation initiatives in the 1970s and 1980s and foresters and academics currently involved in forest conservation in Sierra Leone. Extensive time was also spent in Sierra Leone, which included visits to different historical conservation sites.

After collection, the data was sorted thematically and temporally in a computer database: thematically along the lines of the Forestry Department's four historical major programme areas – reservation (Chapter 2), plantations (Chapter 3), exploitation (Chapter 4), wildlife conservation (Chapter 5) – which make up four of this book's chapters; and temporally in terms of the date of the publication (or the specific dates that it discusses); each database section contained twenty-year intervals (e.g. 1900 to 1919; 1920 to 1939 etc.). This was done in such a manner as to allow for a comprehensive evaluation of the data. The duplication of material that crossed thematic or temporal categories proved to be manageable and even helped to reinforce findings across the different chapters. The data in each section of the database was systemically analysed in thematic and temporal order (i.e. starting with 'reservation: 1990 to 1919' and finishing with 'wildlife conservation 2000 to present'). From this data a historical narrative was developed for each of the book's chapters.

Notes

1. P. Munro, 'Deforestation: Constructing Problems and Solutions on Sierra Leone's Freetown Peninsula', *The Journal of Political Ecology* 16 (2009), 104–24.
2. K.I. Dumbuya, 'Sierra Leone: Environmental Forum – VP Sumana Plants Trees and Lays Pillar for Western Area Peninsular Forest Reserve', *Concord Times* (2011) 13 June.
3. Dumbuya, 'Sierra Leone'.
4. Dumbuya, 'Sierra Leone'.
5. E.K. Alieu, 'Building on Local Foundations: Enhancing Local Community Support for Conservation', *Unasylva* 61 (2010): 22–27.
6. A. Afzelius, *Adam Afzelius Sierra Leone Journal 1795–1796* (Almqvist and Wiksell, 1967); R. Clarke, 'Sketches of the Colony of Sierra Leone and its Inhabitants', *Transactions of the Ethnological Society of London* 2 (1863): 320–63; J.L. Wilson, *Western Africa: Its History, Condition and Prospects* (Sampsons Low, 1856).
7. C.E. Lane-Poole, *Report on Forests of Sierra Leone* (Waterlow & Sons, 1911), 4.
8. B. Braun, *The Intemperate Rainforest: Nature, Culture, and Power on Canada's West Coast* (University of Minnesota Press, 2002).
9. G.H. Endfield, 'Environmental History', in N. Castree, D. Demeritt, D. Liverman and B. Rhoads (eds), *A Companion to Environmental Geography* (Blackwell, 2009), 223–37.
10. S.P. Hays, *Explorations in Environmental History* (University of Pittsburgh Press, 1998).
11. R.P. Neumann, *Making Political Ecology* (Hodder Arnold, 2005); S. Batterbury, 'Political Ecology', in N. Castree, M. Hulme, & J.D. Proctor (eds), *Companion to Environmental Studies* (Routledge, 2018), 439–42.
12. P. Robbins, *Political Ecology*, Vol. 2 (Wiley-Blackwell, 2012).
13. For example, see J. Fairhead and M. Leach, *Reframing Deforestation: Global Analyses and Local Realities with Studies in West Africa* (Routledge, 1998); J. Fairhead and M. Leach, 'Reconsidering the Extent of Deforestation in Twentieth-Century West Africa', *Unasylva* 192 (1998), 38–46; R.P. Neumann, *Imposing Wilderness: Struggles over Livelihood and Nature Preservation in Africa* (University of California, 1998); R.L. Bryant, *The Political Ecology of Forestry in Burma 1824–1994* (Hurst & Company, 1997); N.L. Peluso, *Rich Forests, Poor People: Resources Control and Resistance in Java* (University of California Press, 1992).
14. Neumann, *Making Political Ecology*.
15. Robbins, *Political Ecology*.
16. K.H. Offen, 'Historical Political Ecology: An Introduction', *Historical Geography* 32 (2004), 19–42.
17. N.L. Peluso and P. Vandergeest, 'Genealogies of the Political Forest and Customary Rights in Indonesia, Malaysia, and Thailand', *The Journal of Asian Studies* 60 (2001), 761–812.

18. Neumann, *Making Political Ecology*; I. Vaccaro, O. Beltran and P.A. Paquet, 'Political Ecology and Conservation Policies: Some Theoretical Genealogies', *The Journal of Political Ecology* 20 (2013), 255–72.
19. T. Forsyth, 'Politicising Environmental Explanations: What Can Political Ecology Learn from Sociology and Philosophy of Science?', in M.J. Goldman, P. Nadasdy and M.D. Turner (eds), *Knowing Nature: Conversations at the Intersection of Political Ecology and Science Studies* (University of Chicago Press, 2011), 31–46.
20. For example, Fairhead and Leach, *Reframing Deforestation*; Neumann, *Imposing Wilderness*; Bryant, *The Political Ecology of Forestry*; Peluso, *Rich Forests, Poor People*; P. von Hellermann, *Things Fall Apart?: The Political Ecology of Forest Governance in Southern Nigeria* (Berghahn Books, 2013).
21. D. Weiner, 'A Death-Defying Attempt to Articulate a Coherent Definition of Environmental History', *Environmental History* 10 (2005), 404–20.
22. K. Asdal, 'The Problematic Nature of Nature: The Post-constructivist Challenge to Environmental History', *History and Theory* 42(4) (2003), 60–74.
23. T. Griffiths, 'The Humanities and an Environmentally Sustainable Australia', *Australian Humanities Review* 43 (2007).
24. P.G. Munro and M. Melo Zurita, 'The Role of Cenotes in the Social History of Mexico's Yucatan Peninsula', *Environment and History* 17 (2011), 583–612.
25. G.H. Maddox, 'Africa and Environmental History', *Environmental History* (1999), 162–67.
26. A notable exception is Pauline von Hellermann's excellent monograph *Things Fall Apart? The Political Ecology of Forest Governance in Southern Nigeria.*
27. P.G. Munro and G. van der Horst, 'Breaks with the Past: Conflict, Displacement, Resettlement and the Evolution of Forest Socio-Ecologies in Sierra Leone', in J. Lahai and T. Lyons (eds), *African Frontiers: Insurgency, Governance and Peacebuilding in Post-colonial States* (Ashgate Publishing, 2015), 119–30.
28. D.K. Davis, 'Desert "Wastes" of the Maghreb: Desertification Narratives in French Colonial Environmental History of North Africa', *Cultural Geographies* 11 (2004), 359–87.
29. J. Fairhead and M. Leach, 'False Forest History, Complicit Social Analysis: Rethinking Some West African Environmental Narratives', *World Development* 23 (1995), 1023–35.
30. J. Fairhead and M. Leach, *Misreading the African Landscape: Society and Ecology in a Forest-Savanna Mosaic* (Cambridge University Press, 1996).
31. C.A. Kull, 'Deforestation, Erosion, and Fire: Degradation Myths in the Environmental History of Madagascar', *Environment and History* (2000), 423–50.
32. J.C. McCann, 'The Plow and the Forest: Narratives of Deforestation in Ethiopia, 1840–1992', *Environmental History* 2 (1997), 138–59; also see J.C. McCann, *Green Land, Brown Land, Black Land: An Environmental History of Africa, 1800–1990* (Heinemann, 1999).
33. W. Beinart, 'African History and Environmental History', *African Affairs* 99 (2000), 269–302.

34. For example, see G.A. Barton, *Empire Forestry and the Origins of Environmentalism* (Cambridge University Press, 2002); Richard Grove traces this history even earlier but also sees colonial foresters playing a critical role: R. Grove, *Green Imperialism: Colonial Expansion, Tropical Island Edens and the Origins of Environmentalism, 1600–1860* (Cambridge University Press, 1995).
35. For example, see Fairhead and Leach, *Reframing Deforestation*; Peluso, *Rich Forests, Poor People*; Bryant, *The Political Ecology of Forestry*.
36. J. Martinez-Alier, *The Environmentalism of the Poor: A Study of Ecological Conflicts and Valuation* (Edward Elgar Publishing, 2003).
37. B. Bennett, *Plantations and Protected Areas: A Global History of Forestry Management* (MIT Press, 2015), 46.
38. K. Sivaramakrishnan, *Modern Forests: Statemaking and Environmental Change in Colonial Eastern India* (Stanford University Press, 1999), 243.
39. R.A. Cline-Cole, 'Redefining Forestry Space and Threatening Livelihoods in Colonial Northern Nigeria', in R.A. Cline-Cole and C. Madge (eds), *Contesting Forestry in West Africa* (Ashgate, 2001), 38.
40. M. Crowder, 'Indirect Rule: French and British Style', *Africa* 34(3) (1994), 197–205; K. Good, 'Settler Colonialism: Economic Development and Class Formation', *The Journal of Modern African Studies* 14(4) (1976), 597–620.
41. R. Grove and T. Falola, 'Chiefs, Boundaries, and Sacred Woodlands: Early Nationalism and the Defeat of Colonial Conservationism in the Gold Coast and Nigeria, 1870–1916' *African Economic History* 24 (1996), 1–23.
42. S. Berry, 'Hegemony on a Shoestring: Indirect Rule and Access to Agricultural Land', *Africa* 62 (1992), 327–27.
43. H. Bernstein and P. Woodhouse, 'Telling Environmental Change Like It Is? Reflections on a Study in Sub-Saharan Africa', *Journal of Agrarian Change* 1(2) (2001), 283–324.

CHAPTER 1

Sierra Leone

Colonial Seeds Enter the African Soil

Serra Lyoa

The name Sierra Leone is a semantic corruption of the Portuguese designation *serra lyoa*. It means lion mountain. A popular, yet possibly apocryphal,[1] historical account traces the genesis of the name to a 1492 West African voyage by the Portuguese explorer Pedro da Çintra, who ascribed the epithet to a forested mountainous peninsula that bore a resemblance to the large African feline.[2] Thunderstorms in the area, which roared like lions, potentially gave the name further pertinence.[3] The 'lion mountain' name was later stretched as the moniker for the entire country, with the peninsula being rebranded, in less lyrical terms, as the Western Area Peninsula. This forested mountainous peninsula, some 300 odd years later, would eventually become the founding site for Freetown, Sierra Leone's capital city. It is the same forested mountain that Sierra Leone's Vice President made his Tree Planting trek up in 2010. Prior to the founding of Freetown, however, the *Serra Lyoa* Peninsula was a popular port of call for Portuguese traders: it was a sheltered place to source water and wood, as well as to exchange goods with the local Sherbro and Temne populations.[4] Many of these Portuguese traders permanently settled on the Peninsula, intermarrying with local chiefdom families. These Afro-Portuguese families would become notorious operators within the transatlantic slave trade.[5] With the British colonisation of the broader Sierra Leone territory during the 1890s, the name Sierra Leone was adopted for the entire country – an area of 71,740 km^2 (about the same size as

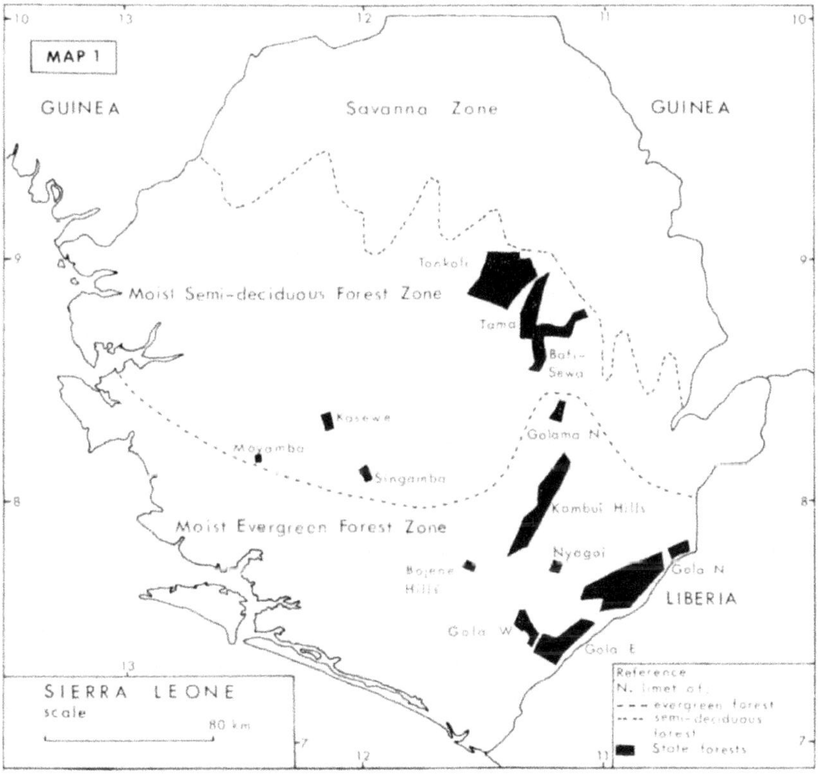

Figure 1.1 Map of vegetation zones in Sierra Leone. Sourced from Peter S. Savill, 'The Composition of Climatic Forest Formations in Sierra Leone', *The Commonwealth Forestry Review* 52(1) (1973), 67–71.

Ireland) – which is home to a large range of flora and fauna due to its diverse climate.

Overall, Sierra Leone's vegetation is highly complex but is generally characterised by a fine-grained matrix of diverse patches – the combined product of soil and interspecies competition as well as other dynamic climatic and anthropogenic factors. This vegetation, following the classification system developed by Peter Savill,[6] falls into three broad geographical zones that traverse Sierra Leone, eponymously named after the forest types that they contain: 1) The Moist Evergreen Forest Zone; 2) The Moist Semi-Deciduous Forest Zone; and 3) The Savanna (Woodland) Zone (see Figure 1.1). The transition between the different zones is gradual; with rainfall being the main influencing variable along the north-south axis on which they are demarcated (i.e. the country gradually becomes drier as one heads north).

The southernmost zone reportedly contains Tropical Moist Rainforest. The annual heavy rainfall (2,750 to 3,000mm per annum) and the high humidity of this zone are conducive to producing closed canopied forests that are numerically rich in tree species.[7] The forests in this area are characterised by an abundance of *Heritiera utilis*[8] and *Lophira alata*,[9] but a range of other species are also commonly found, including *Piptadenisastrum africanum*, *Erythrophleum ivorense*, *Parkia bicolor*, *Klainedoxa gaborensis* and *Parinaria excelsa*.

The middle zone contains Moist Semi-Deciduous Rainforest. Its annual rainfall (between 2,500 and 2,750mm per annum) in conjunction with a short period of relatively low humidity each year (during the dry season) has meant that its forests have a greater proportion of deciduous tree species, less canopy density and less biodiversity in comparison to the forests further south. Savill divides this zone into a further three subcategories; 1) Tonkoli type forests, effectively in the centre of Sierra Leone, where *Terminalia ivorensis and Parkia bicolor* are more prominent; 2) Kasewe type in the south-western part of the country, where the tree species *Nesogordonia papverifera* is abundant. Finally, 3) Forest outlier type, further to the east, which is similar to the Tonkoli type except that the tree species *Triplochiton scheroxylon* is more commonly found.[10]

The last zone crosses the north of the country and contains Woodland Savanna. It has the lowest rainfall of the three zones (less than 2,500mm per annum) as well as the longest dry season, resulting in a less dense and lower canopy height even in mature woodland patches.[11] The most common tree species found in this region is *Lophira lanceolate*.

Mangrove forests are also found in the saline tidal areas below the high tide mark in the estuaries of Sierra Leone's main river and to some extent along the coast. The mangrove species found are *Rhizophora spp*, *Avicennia Africana*, *Conocarpus erectus* and *Laguncularia racemosa*.[12]

Sierra Leone's mammals include the rare and elusive pygmy hippopotamus (*Choeropsis liberiensis*), which is found in the east of the country in the Gola Forest and surrounds, as well as fifteen species of primates, including bushbabies, monkeys and the common chimpanzee (*Pan troglodytes*).[13] The country is particularly renowned for its population of Colobus Monkeys – Olive (*Procolobus verus*); Red (*Procolobus badius*); and Black and White (*Colobus polykomos*) varieties – which have been the focus of numerous primate research projects in the east of Sierra Leone.[14] Sierra Leone is also well known for its diversity of bird species, notably the endemic white-necked rockfowl (*Picathartes gymnocephalus*),[15] which was the focus of a 1950s expedition to Sierra Leone by the famed British Broadcasting Corporation (BBC) naturalist David Attenborough.[16] Findings and footage from the expedition was used in the first pilot episode of *Zoo Quest*, the documentary series that would effectively launch Attenborough into his international stardom. Leopards (*Panthera pardus*) and elephants (*Loxodonta Africana*) were both once relatively common in Sierra Leone, but both are now near extinct in the country due to hunting during the colonial and early independence eras.

Sierra Leone's diverse fauna population was also joined by another mammal some 4,500 to 11,000 years ago: humans.[17] Early on, human population groups would likely have been spread sparsely; however, between 1100CE and 1600CE – around the same time the Portuguese were visiting Sierra Leone's shoreline – there was a mass migration of the Mane people from the neighbouring Guinea region that changed the population dynamic considerably.[18] The impacts of this mass migration on Sierra Leone's forest cover have been the focus of tense academic debate. Endres Nyerges, for example, has suggested that deforestation was considerable during the period, as the Mane population included blacksmith professionals who brought with them the art of charcoal making, a potentially intensive forest-harvesting activity.[19] Anthropologists Melissa Leach and James Fairhead suggest such conclusions are based on a misreading of the landscape and echo flawed colonial narratives of forest cover change.[20] Limited historical data has ultimately meant that much of this debate is at the level of informed speculation.[21] Likewise, little is known about the impact that the slave trade, which devastated the region's populous from the 1500s to the 1700s, had on human-forest dynamics.

These historical debates over forest cover are not just relevant for academic cavils, as ideas about how much of Sierra Leone was 'originally' covered in forest ultimately form an important backstory for contemporary forest conservation efforts. Although Sierra Leone's actual pre-colonial forest cover is effectively unknown due to a lack of reliable data, prominent forest conservation literature often mythologises Sierra Leone as being almost completely covered in thick tropical forests in the not distant, pre-colonial, past – a forest that was decimated due to human influence.[22] It is a mythology that underlies a crisis narrative for urgent action to protect the forest, which, during the colonial and arguably much of the postcolonial era, has generally meant promoting more control over the forests by the central government. Indeed, it was an important narrative underpinning the establishment of colonial forestry in Sierra Leone. Therefore, although the Forest Department in Sierra Leone was set up in 1911, the deeper roots of the rationale behind setting up the Department were arguably the product of a European perspective and political economic approach that had developed and emerged during the previous few centuries.

Sierra Leone's Colonial Forestry Beginnings

The year 1735 provides a seminal date to unpack early colonial forest environmental history. It was a watershed year for the emergence of European 'Green Imperialism' due to the passing of three significant events that aided in the advancement of a naturalist way of thinking and the related empire ambitions. It offers insight into the broad colonial history of colonial forestry imperialism that would occur later on in Sierra Leone and other African colonies. The first 1735 event was a major inter-European scientific expedition launched to 'measure' the shape of the world. It involved sending teams to the Arctic Pole and the Equator in South America. The expedition involved a team of scientists from competing, and often in conflict, nation states in Europe: science was being placed ahead of imperialist nationalism. The findings from the expedition were relatively unremarkable; it was more noteworthy for the popularity of the 'traveller' accounts written by the various expedition members about their struggles in travelling through parts of South America. These were published and read by a wide populace in Europe and helped to pave the way for the 'explorer scientist' vocation.[23] The second event was Carolus Linnaeus's publication of the text *Systema Naturae* (The System of Nature), which led to the Linnaeus System, a classification scheme designed to categorise all plant forms on the planet (a system that is still in use today).[24] Linnaeus had managed to bring two centuries of botanical thinking together into a simple and elegant system of classification.[25] The final 1735 event was the drafting of the first major plans for the botanical gardens in Kew, London. These gardens, for the next couple of centuries, would become an important nexus of botanical, and therefore economic, knowledge, storage and transfer.[26] Overall, these three events precipitated an era of close entwinement between science, commerce and imperial ambitions.

Throughout the eighteenth century, Linnaeus' botanist disciples – dubbed the 'Apostles of Linnaeus' – were sent around the globe to examine and collect plant species. This was the same period in which the modern state of Sierra Leone was being established. In its earliest incarnation, it was a late eighteenth-century experimental attempt, partly initiated by the British Government, to provide a form of redress to former slaves from England, eastern Canada and the Caribbean with the founding of the Freetown settlement in 1792.[27] This was a fledgling settlement that would later be formally annexed by the British in 1808. The design and setting up of the settlement was entangled with broader botanist pursuits, with two of Linnaeus' apostles being sent to Sierra Leone: Andreas Berlin in 1773, and Adam Afzelius from 1792 to 1796.[28] Berlin's tenure was short – marked by drunkenness and then suddenly death from fever (likely caused by malaria) some three months after his arrival.[29] Afzelius' tenure was much longer; he travelled with, and worked for, the Sierra Leone Company, which had been set up to manage the new Freetown settlement. Afzelius encountered much bad luck during his stay; he lost his diary specimens during a French raid on the freed slave colony and more specimens later on during a storm. Nevertheless, the research of his that survived was held in high regard, and the plant genus *Afzelia Sm.* was named in his honour.[30] Fittingly, one of the trees Afzelius identified and had named after him in Sierra Leone, *Afzelius Africana*, would prove to be a valuable tree for producing commercial timber.[31]

The work of Linnaeus' botanists, and indeed botany 'science' at the time, was not an apolitical, economically disinterested affair. Linnaeus requested that students should make economic plants their priority,[32] as it was Linnaeus' belief that nature and the economy were inherently linked: 'Nature has arranged itself in such a way that each country produces something especially useful; the task of economics is to collect [plants] from other places and cultivate [at home] such things that don't want to grow [here] but can grow [here].'[33] Linnaeus' young botanists, therefore, often exchanged knowledge about plants with potential commercial opportunities for free passage on government or commercial ships. As Richard Drayton notes, 'Linnaeus more than any other single figure, turned botany into an economic science.'[34] Arguably, though, it was the Kew Gardens project in London, rather than Linnaeus, that benefited the most from these botanist expeditions. Kew Gardens had grown in stature during the late eighteenth and early nineteenth centuries under the tutelage of Joseph Banks,[35] who had previously worked as a botanist (with two of Linnaeus' apostles) on the First Fleet expedition to Australia in the 1760s. Banks, thanks in part to his political connections, managed to transform the Kew project into a centre of botanical knowledge, with botanists around the world sending Banks their collected samples and notes. In exchange, Banks often provided some form of sponsorship.[36] For example, Banks had close links to Afzelius during the latter's Sierra Leonean expedition: Banks sent Afzelius botanical supplies after his previous supplies had been destroyed by a French Buccaneer attack.[37] The British Empire was thus able to gradually secure information on economic plant species around the world: botany science, imperial conquest and the mercantilist commercial empire were becoming closely intertwined projects.

Scientific endeavours, empire trade and European imperialism continued in concert during the nineteenth century. Drawing upon earlier botanical studies, the period included the gradual emergence of a specific pseudo-scientific imaginary of tropical ecology by Europe. A key feature was what has come to be referred to as 'desiccation theory', which postulated that colonial territories around the world were gradually drying up, chiefly due to what European observers perceived as a rapid loss of forest cover. While the notion of desiccation seems to have first arisen as a feature of French colonial discourse, it soon spread to the British administration, in India and its application to mainland Africa was evident already in the early 1800s,[38] largely through the journal accounts of British missionaries and explorers.[39] As Georgina Endfield and David Nash describe, by this time the 'theory' was already well entrenched among European observers, who selectively accepted aspects of local knowledge that supported their preconceptions of desiccation while ignoring or dismissing those that were contradictory.[40] Moreover, such accounts from the field reinforced the validity of the 'theory', and by the 1860s prominent members of the Royal Geographical Society (RGS) had popularised the idea that the African continent was drying up due to indiscriminate tree felling by 'destructive' local populations.[41] Belief in desiccation became even more acute during the late nineteenth and early twentieth centuries with the ecological notion of 'vegetation climax' – that the 'natural' equilibrium state of tropical regions was dense forest; therefore, any parts of the tropics that were not forested were attributed to historical ecological practices of local inhabitants.[42] This 'theory' was taken a step further in West Africa in the 1930s, after the visit of the influential Forestry Professor Edward Percy Stebbing to the region (including some time spent in Sierra Leone), who claimed that because of the destructive practices of the local population, the Sahara Desert was encroaching south and one day would consume the entire West African region.[43] Thus, by the middle of the twentieth century, local populations in sub-Saharan Africa were attributed with causing a cataclysmic environmental disaster.

In parallel to the rise of these 'new' ecological theories, broad transformations occurred in the European Empire – both geographically and philosophically. During the eighteenth and early nineteenth centuries, Britain and other imperial powers generally adopted a *laissez-faire* approach to empire mercantilism. The British Colonial government's involvement generally ended at securing a colonial territory and establishing an administration, and then it was assumed that colonial market forces and individual initiatives would follow.[44] This was the case in Sierra Leone, where, between 1816 and the 1880s, a highly active trade in timber products was conducted by independent European timber barons, many of whom were former slave traders, with the colonial government based in Freetown providing background support.[45] Most of this harvested timber was transported to the United Kingdom and North America to build ships for the British Royal Navy.[46] Thus, to an extent, private trading companies were fulfilling the Empire's resource extraction needs. The impact that this timber industry had on forest cover in Sierra Leone has been subject to academic debate, with different sources arguing that its impact was widespread,[47] substantial[48] or negligible.[49] As most of this activity was conducted by independent merchants outside of the colonial bookkeeping remit, empirical data on the timber operations is quite limited and unreliable. Although, if this timber industry did have an impact on forest cover, it seems most likely that it would have been restricted to coastal and riparian areas, which were accessible by boat – the main mode of transport at the time. Either way, this timber trade ultimately dwindled at the end of the nineteenth century due to the shift to steel-hulled ships, competition from the thriving palm kernel trade and, as some suggest, perhaps from the exhaustion of the most profitable timber reserves.[50]

Beyond this, by the end of the nineteenth century the British Colonial Government became frustrated with the slow uptake of market activity in many of its colonial territories and therefore opted for a 'constructive imperialism' approach,[51] where the Colonial Office would play a more direct role in building infrastructure (e.g. ports, roads, railways) and establishing resource extraction activities (e.g. with Forestry Departments, Mining Departments etc.). A transition most markedly illustrated with the 1890s 'Scramble for Africa', where almost the entire African continent was carved up by half a dozen European powers. In 1896, as a part of this 'scramble', the British Government annexed the Sierra Leonean Protectorate, dramatically expanding Sierra Leone's boundaries from the small peninsula settlement into the hinterland along with what are essentially the current modern-day boundaries of the country – a 120-fold increase in land area. The local population resisted this colonial expansion, most notably with an armed rebellion in 1898 known as the Hut Tax War; however, the colonial invaders were eventually able to establish formal military control after several months of resistance fighting in the north of the Protectorate by Sierra Leoneans, led by their resistance leader Bai Bureh.[52]

The British Colonial Government, perhaps being cautious in the wake of early resistance, opted for an indirect 'Protectorate' governance system for most of Sierra Leone, whereby local Paramount Chiefs were assigned the authority to collect taxes on behalf of the British Government. The Paramount Chiefs also legally remained the official owners of their land. This meant that the colonial presence in Sierra Leone, outside of Freetown and the Western Area Peninsula, which was formerly designated as crown land, was relatively limited. In many ways, therefore, the overall motive for this colonisation in Sierra Leone, and in much of Africa, was less about acquiring territory per se but rather to establish control over (and increase) the extraction of valuable commodities for the benefit of the imperial core.[53] The formalisation and institutionalisation of natural resource exploitation for export was therefore a key focus of the British Colonial Government during the early 1900s, pursued through the establishment of managing bureaucracies for agriculture, mining and forestry.[54] It is in this context that the British Colonial Forestry programmes in Sierra Leone and other West African colonies arose. The British forestry experiment started in its South Asian colonies (India and Burma) and was based largely on a cameralist[55] German model of forestry that had been developed during the 1800s. This German model of forestry was premised on simplifying and rationalising forest resources, so they could easily be exploited. This involved separating forests from people by creating Forest Reserves, changing the ecological make-up of forests so they would produce more economically valuable species (through afforestation and regeneration programmes) and then exploiting these forest resources using available technology. Basically, the German approach was to manufacture a forest that was easier for foresters to count, manipulate, measure, assess and hence exploit.[56]

Critical for implementing this 'forestry science' approach was the training of foresters, and indeed during the nineteenth century there was a distinct transformation of forestry scientists as being 'gentleman amateurs' at the beginning of the century to being a full-time professional class by the end of the century.[57] Young British elites were being sent to Forestry Schools in France and Germany to learn 'forestry science', as forestry schools were not properly established in Britain until the early twentieth century, with the setting up of schools at the University of Edinburgh (Scotland), Oxford University (England) and Bangor University (Wales).[58] Forestry, at the time, was essentially viewed as a nomothetic science – foresters learnt the 'truth' about forests and their management. As a part of this trend, the botanical gardens at Kew also came into their apogee during late nineteenth century – smaller botanical gardens were set up in all of the British Empire's colonies (including one in Sierra Leone in 1895), with Kew Gardens then acting as a metropole disseminating economic plants to botanical gardens around the world so that they could be used in plantation programmes. Overall, the 'practical' experience of forestry in India and Burma during the late nineteenth century was extolled as a model for the rest of the British Empire's colonies and dominions, such as Sierra Leone.[59]

Overall, at the start of the twentieth century, when forestry departments were being set up in Sierra Leone and the rest of Africa, there were two influential ideas underpinning colonial forestry discourses that had emerged during the previous centuries. First, that Africa's forests were under immediate danger of being destroyed due to the activities of the local population (i.e. desiccation theory); and second, with the rise of Colonial Forestry, the British Colonial government believed that they had the 'true' knowledge of how best to manage these forests. Desiccation had become a foundational concept of modern tropical forestry,[60] and European control of the continent's forests was justified within the moral-economic framework of colonisation.[61] The taking control of forests in Africa was framed as being a beneficial gesture to protect the local 'ignorant' population from themselves. Forestry science was seen to provide an unquestionable truth. Louis Lavauden, a prominent French colonial forester in Africa during the first half of the twentieth century, provided a curt recommendation on how colonial foresters should deal with the local population:

> There is a common belief that there is much to be learnt from the native, and the author has heard this principle upheld by many colonials. It is a false doctrine. The competent man has nothing to learn from the native: on the contrary the native has everything to learn from him. And the extent to which incompetents are listened to is, to put it mildly, lamentable.[62]

Lavauden was a graduate of the French National School of Forestry (*École nationale des eaux et forêts*) in Nancy, France. One of his fellow students at the school was the Englishman Charles Lane-Poole, who would later establish and head the Sierra Leone Forestry Department. Lane-Poole shared a similar philosophy to Lavauden. Indeed, Lavauden's ideas were pervasive across forestry departments in Africa for many years, and William Robertson, who headed Sierra Leone's Forestry Department between 1936 and 1939, celebrated the translation of Lavauden's texts into English and drew upon them to help inform his management of the Department.[63]

The above supports Roderick Neumann's conclusion that 'colonial science, with little exception, condemned African land and natural resource use and practices as wasteful, environmentally destructive and inefficient, thereby supporting the state's moral justification for its proprietary claims'.[64] Such crisis narratives were used as justification for experts to become stakeholders, and this, in part, explains the stickiness of some crisis narratives such as desiccation and desertification, which still persist even today, despite empirical evidence indicating that they are based on flawed premises.[65] They persist precisely because they are linked to particular framings of political power. As such, colonialism in Africa should not just be viewed as a project of military occupation and control but rather as one having important discursive elements.[66] With colonialism came new ideas about nature: 'Nature was there to be disciplined and regulated, harnessed to the imperative of the imperial.'[67] The engagements between coloniser and colonised, between metropolitan and peripheral economies, and between modern technology and nature, were not only direct and material but also discursive. Knowledge of the colonised world and its increasingly transformed nature were intrinsic to colonial domination.[68]

Forestry Department

To realise these objectives, a professional forestry management body had to be set up in each colony. This process was started in Sierra Leone by the colonial government in 1908, with the sending of the prominent colonial forester Arthur Unwin from his posting in Nigeria[69] to evaluate Sierra Leone's forestry potential.[70] Unwin, drawing upon the assumption that the country was once heavily forested, gave a damning account of the Sierra Leonean forest landscape, describing the country as being heavily deforested and placing most of the blame for this on the local population and their 'wasteful' local methods of forest exploitation.[71] Unsurprisingly, in his final report, Unwin recommended that a Forestry Department should be created, and during his one-month trip across Sierra Leone he noted a number of areas to be converted into Forest Reserves. He also noted that there was some immediate potential for profitable commercial forestry.[72]

In response to Unwin's findings, Charles Lane-Poole was sent to Sierra Leone in 1910 and was tasked with conducting a second evaluation of the country's forest stocks as well as setting up the Forestry Department. He was Sierra Leone's first formal forestry officer:

> I am directed by the Earl of Crewe to inform you [Lane-Poole] that, subject to your being passed as physically fit for service, he proposes to select you for appointment as Forestry Officer in Sierra Leone ... There is at present no Forestry Department in Sierra Leone, and you would be required to undertake the organization of the new Department.[73]

Lane-Poole might seem a curious choice for the task – at the time he was only 26 years old, had never previously seen a tropical forest and had limited forestry-working experience. However, as John Dargavel notes in his biography of Lane-Poole, such situations were not all that uncommon on the colonial frontier, where young, inexperienced officers were often given extraordinary responsibilities.[74] Lane-Poole was certainly not lacking in confidence, an 'empire man' with an aristocratic background and French training in forestry, he had a resolute assurance in his own abilities and a fervent dedication to the 'science' of forestry. Dargavel dubbed him 'the Zealous Conservator'[75] and provides a compelling description of the fervent colonial forester:

> Charles Edward Lane-Poole commanded attention. He was a little over average height with a square face, a strong jaw and a determined eye. Old photographs show him serious, looking straight ahead with an energy about him. He was well-dressed, shoes polished, trousers pressed, tie straight and a hook in the place of his left hand … He was captured by the ideals of forest conservation and its science, and he followed their dictates across the world, irrespective of personal consequences or political reality, through much of the twentieth century. He was truly a zealous Conservator of Forests, so sure in his path that he was intolerant of any other views.[76]

Arguably, at least during the early colonial period, Lane-Poole would be the one individual who had the greatest impact in shaping formal forestry in Sierra Leone.

Like Unwin, Lane-Poole embarked on a relatively rapid (three month) tour of Sierra Leone and came to the same conclusion about the state of the country's landscape – that its forests had almost been completely destroyed by the destructive activities of the local populace, declaring that: 'the Rain Forests at one time must have covered the whole country …, 99 per cent of this [forest] has been destroyed.'[77] Lane-Poole's vision for the future of forestry in Sierra Leone differed somewhat from Unwin. While he did not disagree with the need to establish a commercial forestry sector in Sierra Leone, he ultimately saw it as a much more distant prospect. Before it was to be achieved, the country needed to set up a large forest estate, which involved the local population relinquishing control of their forests over to a Forestry Department.

Unwin and Lane-Poole's claims of forest destruction echoed the broader colonial (and racist) thorough-going distrust of native environmental competence. Lane-Poole, in particular, understood forestry as being an elite science and that the *only* way to learn how to manage forests correctly was through training at a respected forestry school in Europe. As such, throughout his career Lane-Poole had form in admonishing European forestry colleagues who he felt had not trained at an exclusive-enough forestry school.[78] It was therefore inconceivable in Lane-Poole's narrow, elitist, European perspective that the local Sierra Leonean population could have competent forms of forest management.

This was and still is a problematic perspective. Not least because Lane-Poole, among many other European colonial foresters, drew their conclusions from a systematic misreading of the Sierra Leonean landscape through the lenses of pseudo-scientific ecological notions such as 'succession', 'equilibrium' and (deforestation-induced) desiccation[79]– notions that led them to assume that Sierra Leone was once heavily covered in forest, an assertion that is not supported by any definitive evidence. Their assumptions about historical forest cover had more to do with their European biases than understandings of African forested landscapes. Lane-Poole's misreading is perhaps unsurprising, given the mix of his inexperience, his overconfidence and that his visit to Sierra Leone was the first time he had ever seen a tropical forest (his previous experience had been three years of work in the temperate forests of South Africa).[80] It is a misreading, nevertheless, that still resonates in contemporary environmental discourses in Sierra Leone; as previously mentioned, the country is popularly mythologised as once having a heavily forested past sometime in the pre-colonial era.[81]

After his evaluation of the country's forests, Lane-Poole drafted a Forestry Law and established a Sierra Leonean Forestry Department. The forest law, known as *An Ordinance for the Regulation of Forest Reserves in the Colony and Protectorate* (or *Forest Ordinance* for short) was formerly passed in 1912; much of its text, however, had been copied and pasted from the Nigerian Forestry Ordinance.[82] The law contained information on how Forest Reserves were to be formally established, and it provided some rules regarding timber extraction.

Since its foundation in 1911, the Sierra Leonean Forestry Department's formal structure has remained relatively static despite its jurisdiction being constantly shifted under different government bodies (e.g. Ministry of Natural Resources; Ministry Of Agriculture and Forestry; National Commission for Environment and Forestry; the Ministry of Agriculture, Forestry and Food Security and Local Governments) along with constant name changes (e.g. the Forestry Division, the Forest Department; Lands and Forests Department). Operating with a hierarchical structure, the Forestry Department was made up of self-described Senior Staff, Intermediate Staff and Junior Staff.

The Senior Staff included a formal Director of Forestry (known as The Conservator of Forests (1911 to 1951); the Chief Conservator of Forests (1952 to 1990s); and the Director of Forests (since the 1990s)) and a number of Assistant Directors (known as Assistant Conservators of Forests (1911 to 1951); Conservators of Forests (1952 to 1990s)). For most of the colonial era, these staff were European. Intermediate Staff included District Forest Officers and Forest Supervisors. Again, in the first years of the Department, these staff were often European; however, by the 1940s these roles were filled by Sierra Leoneans. Finally, the Junior Staff were Forest Rangers (Chief, 1st and 2nd grade), Foresters and Forest Guards, who were all mainly based in the field. The junior staff were always Sierra Leonean. The Forestry Department, thus, during the colonial era was run by an expatriate colonial elite, with Sierra Leoneans filling lower-rung posts. Only Europeans were given the title of Conservators of Forests.[83]

Each Assistant Director was in charge of a subsection of the Department's operations. During the colonial era, there were three formal areas of operation: Reservation, Afforestation and Utilisation (Exploitation). In the postcolonial era, these have changed to Commercial Forestry, Conservation & Wildlife Management and Community Forestry. Often these Assistant Directors also have their own deputies. Below the Assistant Directors were the District Forestry Officers, who oversaw operations in their respective districts. There are currently thirteen in total (one for each district); however, during the colonial era there were usually only three of four in operation at one time due to an overall lack of funds. Below these District Forestry Officers were the field staff, such as Rangers, Foresters and Forest Guards.[84]

After founding the Forestry Department, Lane-Poole became Sierra Leone's first Conservator of Forests, and he headed the Department for its first five years until 1916. Given this, for its first fifteen years (and with a later semi-revival during the 1940s and 1950s) the Forestry Department followed a vision close to Lane-Poole's, with an almost relentless focus on establishing forest reserves. They were seen as the answer to addressing the perceived forest destruction. As Lane-Poole articulated in his report, 'I have pointed out the necessity of protecting the forest … and the *only way this can be done* effectively is by the formation of Forest Reserves.'[85] Indeed, John Dargavel's biography of Lane-Poole summed up his time in Sierra Leone with the one word chapter title 'demarcation'. Forestry in Sierra Leone during Lane-Poole's time was all about demarcating Forest Reserves.[86]

Notes

1. C. Magbaily Fyle 'Pedro da çintra did Not Name Sierra Leone: An Exploration into Available Evidence', *Concord Times*, June 22, 2015.
2. J. Alie, *A New History of Sierra Leone* (MacMillian, 1990); J. Kaifala, *Free Slaves, Freetown, and the Sierra Leonean Civil War*. African Histories and Modernities (Palgrave Macmillan, 2017), 1–8.

3. Alie, *A New History of Sierra Leone*.
4. C. Fyfe, *A History of Sierra Leone* (Oxford University Press, 1962).
5. W. Rodney, 'African Slavery and Other Forms of Social Oppression on the Upper Guinea Coast in the Context of the Atlantic Slave-Trade', *The Journal of African History* 7(3) (1966), 431–43.
6. P.S. Savill, *The Classification of the Forest of Sierra Leone, with a Field Key for the Most Common Genera and Species* (University of Bangor, 1969); P.S. Savill, 'The Composition of Climatic Forest Formations in Sierra Leone', *The Commonwealth Forestry Review* 52(1) (1973), 67–71.
7. Savill, 'The Composition of Climatic Forest Formations', 67–71.
8. In terms of local Sierra Leonean languages *Heritiera utilis* is known as *Ka-fɔlfɔl* in Temne; *Hamɔn* in Krio; *Yawii* in Mende; and *Denɛrɛnafa* in Kuranko.
9. In terms of local Sierra Leonean languages, *Lophira alata* is known as ɛ-Nanka in Temne, Ku-Mannaka in Limba; and Njombo-wuli in Mende.
10. Savill, 'The Composition of Climatic Forest Formations', 67–71.
11. Savill, *The Classification of the Forest of Sierra Leone*.
12. Savill, *The Classification of the Forest of Sierra Leone*.
13. G.H. Whitesides, J.F. Oates, S.M. Green and R.P. Kluberdanz, 'Estimating Primate Densities from Transects in a West African Rain Forest: A Comparison of Techniques', *The Journal of Animal Ecology* 57(2) (1988), 345–67.
14. Whitesides et al., 'Estimating Primate Densities', 345–67; J.F. Oates, 'The Diet of the Olive Colobus Monkey, Procolobus Verus, in Sierra Leone', *International Journal of Primatology* 9 (1988), 457–78; J.F. Oates and G.H. Whitesides, 'Association Between Olive Colobus (Procolobus Verus), Diana Guenons (Cercopithecus Diana), and Other Forest Monkeys in Sierra Leone', *American Journal of Primatology* 21(2) (1990), 129–46.
15. H.S. Thompson, 'Status of White-Necked Picathartes – Another Reason for the Conservation of the Peninsula Forest, Sierra Leone', *Oryx* 27(3) (1993), 155–58.
16. D. Attenborough, 'Expedition to Sierra Leone', *Zoo Life* 10 (1955), 11–20.
17. M.I. Bird and J.A. Cali, 'A Million-Year Record of Fre in Sub-Saharan Africa', *Nature* 394 (1988), 767–69.
18. G.E. Brooks, *Landlords and Strangers: Ecology, Society, and Trade in Western Africa, 1000–1630* (Westview Press, 1993); R. Walter, 'A Reconsideration of the Mane Invasions of Sierra Leone', *The Journal of African History* 8(2) (1967), 219–46.
19. A.E. Nyerges, 'Ethnography in the Reconstruction of African Land Use Histories: A Sierra Leone Example', *Africa* 66 (1996), 122–44.
20. J. Fairhead and M. Leach, *Reframing Deforestation: Global Analyses and Local Realities with Studies in West Africa* (Routledge, 1998); J. Fairhead and M. Leach, 'Shaping Socio-ecological and Historical Knowledge of Deforestation in Sierra Leone, Liberia and Togo', in R.A. Cline-Cole and C. Madge (eds), *Contesting Forestry in West Africa* (Ashgate, 2000), 64–95.

21. P.G. Munro and G. van der Horst, 'Contesting African Landscapes: A Critical Reappraisal of Sierra Leone's Competing Forest Cover Histories', *Environment and Planning D* 24(4) (2016), 706–24.
22. For an overview of the debates, see Munro and Van der Horst, 'Contesting African Landscapes, 706–24.
23. M.L. Pratt, *Imperial Eyes: Travel Writing and Transcultration* (Routledge, 1992).
24. C. Linnaeus, *Systema Naturae* (Regnum Animale, 1735).
25. Pratt, *Imperial Eyes.*
26. R. Drayton, *Nature's Government: Science, Imperial Britain, and the 'Improvement' of the World* (Yale University Press, 2000).
27. S. J. Braidwood, *Black Poor and White Philanthropists: London's Blacks and the Foundation of the Sierra Leone Settlement 1786–1791* (Liverpool University Press, 1994).
28. Braidwood, *Black Poor and White Philanthropists*; S. Douglas, 'The Making of Scientific Knowledge in an Age of Slavery: Henry Smeathman, Sierra Leone and Natural History', *Journal of Colonialism and Colonial History* 9 (2008); L. Koerner, *Linnaeus: Nature and Nation* (Harvard University Press, 1999).
29. Douglas, 'The Making of Scientific Knowledge'.
30. Some of Afzelius's diaries from the period are still available: see A. Afzelius, *Adam Afzelius Sierra Leone Journal 1795–1796* (Almqvist und Wiksell, 1967).
31. D. G. Thomas, 'Report: The Forest Authority, Sierra Leone', paper presented at The British Empire Conference (London, 1924); P.G. Munro and G.A. Hiemstra-van der Horst, 'Conserving Exploitation?: A Political Ecology of Forestry Policy in Sierra Leone', *The Australasian Review of African Studies* 32(1) (2011), 59–72.
32. Drayton, *Nature's Government.*
33. Linnaeus cited in Koerner, *Linnaeus: Nature and Nation.*
34. Drayton, *Nature's Government.*
35. Patrick O'Brian provides a well written and detailed overview of Joseph Bank's life: see Patrick O'Brian, *Joseph Banks: A Life* (University of Chicago Press, 1997).
36. Drayton, *Nature's Government.*
37. C. Eagleton, 'Sara Sophia Banks, Adam Afzelius and a Coin from Sierra Leone', in A. Craciun and S. Schaffer (eds), *The Material Cultures of Enlightenment Arts and Sciences* (Palgrave Macmillan, 2016), 203–6.
38. R.H. Grove, 'Colonial Conservation, Ecological Hegemony and Popular Resistances: Towards a Global Synthesis', in J.M. MacKenzie (ed.), *Imperialism and the Natural World* (Manchester University Press, 1990), 15–50.
39. G. H. Endfield and D.J. Nash, 'Missionaries and Morals: Climatic Discourse in Nineteenth-Century Central Southern Africa', *Annals of the Association of American Geographers* 92 (2002), 727–42.
40. G.H. Endfield and D.J. Nash, 'Drought, Desiccation and Discourse: Missionary Correspondence and Nineteenth-Century Climate Change in Central Southern Africa', *The Geographical Journal* 168 (2002), 33–47.

41. Endfield and Nash, 'Drought, Desiccation and Discourse', 33–47; R.H. Grove, 'Imperialism and the Discourse of Desiccation: The Case of Global Environmental Concerns and the Role of the Royal Geographical Society, 1860–1880', in R.B.M. Bell and M. Heffernan (eds), *Geography and Imperialism, 1820–1940* (Manchester University Press, 1995), 36–53; R.H. Grove, *Green Imperialism: Colonial Expansion, Tropical Island Edens and the Origins of Environmentalism, 1600–1860* (Cambridge University Press, 1996).
42. F.E. Clements, *Plant Succession: An Analysis of the Development of Vegetation* (Carnegie Institute, 1919); F.E. Clements, 'Nature and Structure of the Climax', *Journal of Ecology* 24 (1936), 252–84; G.P. Marsh, *Man and Nature; or Physical Geography as Modified by Human Action* (Charles Scribner, 1864).
43. E.P. Stebbing, 'The Encroaching Sahara: The Threat to the West African Colonies', *Geographical Journal* 85 (1936), 506–19; E.P. Stebbing, *The Forests of West Africa and the Sahara: A Study of Modern Conditions* (W & R Chamber, 1937): 255; E.P. Stebbing, 'The Threat of the Sahara', *Journal of the Royal African Society* 36 (1937), 3–35; E.P. Stebbing, 'The Man-Made Desert in Africa: Erosion and Drought', *Journal of the Royal African Society* 37 (1938), 3–40.
44. M. Worboys, 'The Imperial Institute: The State and the Development of the Natural Resources of the Colonial Empire, 1887–1923', in J.M. MacKenzie (ed.), *Imperialism in the Natural World* (Manchester University Press, 1990), 164–86.
45. F.A. Akiwumi, 'Conflict Timber, Conflict Diamonds: Parallels in the Political Ecology of 19th and 20th Century Resource Exploitation in Sierra Leone', in K. Konadu-Agyemang (ed.), *Africa's Development in the Twenty-first Century: Pertinent Socio-economic and Development Issues* (Ashgate, 2006), 109–25. F.A. Akiwumi, 'Environmental and Social Change in Southwestern Sierra Leone: Timber Extraction (1832–1898) and Rutile Mining (1967–2005)', PhD Thesis (Texas State University, 2006).
46. Fyfe, *A History of Sierra Leone*.
47. D.C. Dorward and A.I. Payne, 'Deforestation, the Decline of the Horse, and the Spread of the Tsetse Fly and Trypanosomiasis (nagana) in Nineteenth Century Sierra Leone', *The Journal of African History* 16(2) (1975), 239–56.
48. A.C. Millington, 'Environmental Degradation, Soil Conservation and Agricultural Policies in Sierra Leone, 1895–1984', in D. Anderson and R. Grove (eds), *Conservation in Africa: People, Policies and Practice* (Cambridge University Press, 1987), 229–48; A.C. Millington, 'Environmental Degradation, Soil Conservation and Agricultural Policies in Sierra Leone, 1895– 1984', *International Journal of Sierra Leone Studies* 1 (1988), 5–25.
49. Fairhead and Leach, *Reframing Deforestation*.
50. Akiwumi, 'Conflict Timber, Conflict Diamonds', 109–25; Akiwumi, 'Environmental and Social Change in Southwestern Sierra Leone'.
51. Worboys, 'The Imperial Institute', 164–86.

52. A. Abraham, 'Nyagua, The British, and the Hut Tax War', *The International Journal of African Historical Studies* 5(1) (1972), 94–98; A. Abraham, 'Bai Bureh, the British, and the Hut Tax War', *The International Journal of African Historical Studies* 7(1) (1974), 99–106; M. Crowder, *West Africa under Colonial Rule* (Hutchinson, 1970).
53. Munro and Hiemstra-van der Horst, 'Conserving Exploitation?', 59–72.
54. P.G. Munro and G. van der Horst, 'Breaks with the Past: Conflict, Displacement, Resettlement and the Evolution of Forest Socio-ecologies in Sierra Leone', in J. Lahai and T. Lyons (eds), *African Frontiers: Insurgency, Governance and Peacebuilding in Post-colonial States* (Ashgate, 2015), Chapter 9: 119–30; P.G. Munro and M. Melo Zurita, 'Aprendiendo del pasado: Impactos históricos de intervenciones en la agricultura de Sierra Leona', *Ecología Política* 38 (2009), 92–95.
55. Cameralism is a German philosophy of public administration. It favoured the strong intervention of the Government into the economy and society to achieve economic outputs.
56. R.P. Neumann, *Making Political Ecology* (Hodder Arnold, 2005).
57. G.A. Barton, *Empire Forestry and the Origins of Environmentalism* (Cambridge University Press, 2002); Drayton, *Nature's Government*; J.M. MacKenzie, 'Introduction', in J.M. MacKenzie (ed.), *Imperialism in the Natural World* (Manchester University Press, 1990), 1–14.
58. *Nature,* The Forestry Department of Edinburgh University 106 (1921), 706–7; J. Burley, R. A. Mills, R.A. Plumptre, P.S. Savill, P.J. Wood and H.L. Wright, 'A History of Forestry at Oxford University', *British Scholar* 1 (2009), 236–61; W. Linnard, 'The History of Forests and Forestry in Wales up to the Formation of the Forestry Commission', PhD thesis (Prifysgol Bangor University, 1979).
59. Barton, *Empire Forestry.*
60. J. Fairhead and M. Leach, 'Desiccation and Domination: Science and Struggles over Environment and Development in Colonial Guinea', *Journal of African History* 41 (2000), 35–54.
61. Endfield and Nash, 'Drought, Desiccation and Discourse', 33–47.
62. L. Lavauden, 'Supplement: The Equatorial Forest of Africa: Its Past, Present and Future', *Journal of the Royal African Society* 36(143) (1937), 3–25.
63. W.M. Robertson, 'Report on Forest Conference in Nigeria in 1937: SLNA: A/71/37, 1937'; W.M. Robertson, 'Sierra Leone Annual Report of the Forest Department for the Year 1936' (Freetown: Government Printer, 1937).
64. R.P. Neumann, 'Nature-State-Territory: Towards a Critical Theorization of Conservation Enclosures', in R. Peet and M. Watts (eds), *Liberation Ecologies: Environment, Development, Social Movements,* 2nd ed. (Routledge, 2004), 195–218.
65. E.M. Roe, 'Except-Africa: Postscript to a Special Section on Development Narratives', *World Development* 23(1995), 1065–69; Also see J. Swift, 'Desertification: Narratives, Winners & Losers', in M. Leach and R. Mearns

(eds), *The Lie of the Land: Challenging Received Wisdom on the African Environment* (James Curry, 1996), 73–90.
66. E. Said. *Orientalism* (Pantheon, 1978).
67. W.M. Adams, 'Nature and the Colonial Mind', in W.M. Adams and M. Mulligan (eds), *Decolonizing Nature: Strategies for Conservation in a Post-colonial Era* (Earthscan, 2002), 15–50.
68. W.M. Adams and M. Mulligan, 'Introduction', in W.M. Adams and M. Mulligan (eds), *Decolonizing Nature: Strategies for Conservation in a Post-colonial Era* (Earthscan, 2002), 1–14.
69. The Nigerian Forestry Department had been set up in 1900. It was the first to be set up in West Africa.
70. A.H. Unwin, *Report on the Forest and Forestry Problems in Sierra Leone* (Waterlow and Sons Limited, 1909). A.H. Unwin, *West African Forests and Forestry* (T. Fisher Unwin Ltd, 1920).
71. Unwin, *Report on the Forest and Forestry Problems*, 33.
72. Unwin, *Report on the Forest and Forestry* Problems, 33.
73. Under Secretary of State (1910). Letter from the Colonial Office to Charles Lane Poole – 2 August. NLA - 3799/5/1.
74. J. Dargavel, *The Zealous Conservator: The Life of Charles Lane Poole* (University of Western Australia, 2008).
75. Dargavel, *The Zealous Conservator*.
76. Dargavel, *The Zealous Conservator*, xi.
77. C.E. Lane-Poole, *Report on Forests of Sierra Leone* (London: Waterlow & Sons, 1911), 6.
78. Dargavel, *The Zealous Conservator.*
79. Fairhead and Leach, *Reframing Deforestation*; Munro and Hiemstra-van der Horst, 'Conserving Exploitation?', 59–72.
80. Dargavel, *The Zealous Conservator.*
81. Munro and Van der Horst, 'Contesting African Landscapes, 706–24.
82. P. Munro, 'A Critical History of Forest Conservation in Sierra Leone', PhD Thesis (University of Melbourne, 2015).
83. Munro, 'A Critical History of Forest Conservation'.
84. Munro, 'A Critical History of Forest Conservation'.
85. Lane-Poole, *Report on Forests of Sierra Leone*, 23.
86. Dargavel, *The Zealous Conservator.*

CHAPTER 2

Reservationism

The forested slopes of the Kambui Hills rise sharply up behind the urban centre of Kenema, providing a dramatic backdrop to Sierra Leone's Eastern Province capital. The Kambui Hills are part of Sierra Leone's Moist Evergreen Forest Zone, with the forest distinct from many of the other forests in the zone due to the prevalence of tree species such as *Brachystegia leonensis Xylia evansii* and *Uapaca spp*, the latter being a popular species for timber harvesting. Early on, the colonial gaze of European foresters fixated on these hills, observing that their forests contained 'quantities of magnificent timber, mahogany and other hard woods'[1] and therefore, even before the colonial Forestry Department was formally established, the colony's first forestry officer, Charles Lane-Poole, declared that 'these forests should be exploited commercially'.[2] The forest's proximity to a major urban centre and the existent railway line added further weight to this case. In 1914, only three years after the establishment of the Forestry Department, plans were made to demarcate the Kambui Hills as a Forest Reserve, the formal establishment of a Reserve being a necessary legal prerequisite for the Forestry Department before it could engage in any exploitation activities in the hills. Lane-Poole had initially assumed this would be a straightforward task; while interviewing Chiefs during his 1910 tour of Sierra Leone, he received positive responses when he ventured the idea of Forest Reserves and therefore he assumed that there would be no 'opposition on the part of the native community' for their establishment.[3] He was wrong.

To establish a Forest Reserve according to the law, the Forestry Department needed to obtain a formal request from all of the Paramount

Chiefs whose land would be affected by the Reserve's establishment. The Kambui Hills was a potentially complicated undertaking in this regard, as its land area crossed multiple chiefdom boundaries, with a large section of it in the Nongowa Chiefdom, whose Paramount Chief, Madam Humonya, was one of the most prominent Paramount Chiefs at the time and held great influence over neighbouring chiefdoms.[4] The process initially seemed straightforward. The District Commissioner, W.D. Bowden, who was based in the nearby urban centre of Kenema, was a great supporter of the Reserve's creation,[5] while the Paramount Chief Madam Humonya, in line with legislation, formally requested for a Forest Reserve to be made out of a 'portion of the Kambui Hills falling within her Chiefdom'.[6] Other Paramount Chiefs in the area fell in line with Humonya, also offering up parts of their chiefdoms for the Forest Reserve, and a slow start was made with the Reserve's demarcation.[7]

After this positive outset, progress towards creating the Reserve rapidly deteriorated; the catalyst being a breakdown of the relationship between Lane-Poole and Bowden on one side, and Paramount Chief Humonya on the other. An argument first broke out between Lane-Poole and Humonya in late 1914. The reason for the argument is unclear, but its occurrence is unsurprising given the detonative personalities of the two protagonists. Lane-Poole (see Figure 2.2), with his irascible personality, had a history of offending those whom he worked with,[8] while Humonya was infamous for her ruthless and uncompromising nature.[9] As Lynda Day details:

> [Humonya was well known for] her excessive demands on [her people's] labour and money. She is said to have demanded one hundred men and one hundred women from each of the different sections in the chiefdom to work her farms in Kenema … when women became pregnant they were kept at work until time for their confinement, men were often punished for misdemeanours, such as gambling, by being sent to prison for months or being put into stocks.[10]

Lane-Poole recorded the encounter with the 'ruddy chief' in his journals as an occasion where 'many winged words were spoke[n] on either side'.[11] Events took a turn for the worse the following year, 1915, when Lane-Poole returned to Kenema to check up on the Reserve's progress, finding out that Humonya had started to develop a farm

Figure 2.1 Passport photograph of Charles Lane-Poole taken in 1915 during his time in Freetown. Source: TNA FO 655/1212 Charles Edward Lane-Poole Passport Photo. Place of Issue: FREETOWN. Name: Charles Edward Lane-Poole. The British National Archives.

right in the middle of the Reserve area. Lane-Poole formally met up with the 'old swine' Humonya, lecturing to her in a manner 'as rude as [he] knew how'.[12] He then wrote to District Commissioner Bowden complaining about the situation and demanding that it should be rectified. It was to no avail, and in 1916 Lane-Poole decided to resign from his post in Sierra Leone – due to frequent illness, general frustration with the state of forestry in the country and the separation from his family. He subsequently took up a forestry position in Western Australia.[13] Humonya remained the Paramount Chief of Nongowa and continued to block the gazettement of the Kambui Hills Forest Reserve for years to come.

After Lane-Poole left Sierra Leone, District Commissioner Bowden brought a number of investigations against Humonya in relation to her 'tyrannical and abusive' style of ruling over her chiefdom subjects. However, he was outmanoeuvred, with Humonya travelling to Freetown and appealing to the Colony's Governor. Remarkably, the Governor sided with Humonya and had Bowden relocated to a different district. Bowden's successor also tried to bring charges against Humonya but failed as well. Humonya was able to stay in office due to the 'close ties with the colonial government' in Freetown that she maintained. Indeed, she stayed in power 'in spite of six prominent men of the chiefdom levying charges against her and two commissions of inquiry that found her guilty of long-standing abuses of power'.[14] It was not until the very end of 1918 that Humonya was finally removed from her position of Paramount Chief. This, however, was not done by the Colonial Government but rather by a revolt from her own subjects, who had tired of her tyrannical rule.[15] This finally paved the way for the formal creation of the Forest Reserve, and just over a year after Humonya was deposed, the Kambui Hills Forest Reserve was formally gazetted in 1920.[16]

The conflict and contestations surrounding the gazettement of Kambui Hills provide empirical and allegorical insights that speak to a core argument in this book: that the establishment of colonial environmental hegemony in Africa was a relatively weak and heavily challenged process. While the Forestry Department, *eventually*, was able to have its Forest Reserve gazetted, a black, African woman, albeit one in a relative position of power, was constantly able to thwart and outmanoeuvre two white, aristocratic, colonial males. It is a story that disrupts some sensibilities about the colonial project; about popular notions of gender relations and ideas surrounding European power and influence. It is a striking example of the dialectics of colonial power in Africa. Britain might have successfully established military control over the geopolitical territory of Sierra Leone in 1896; however, this did not translate into absolute power over its people and land, forests included. Rather, the colonisation project was an ongoing one, a protracted negotiation across different landscapes and with different people in different contexts. The Kambui Hills reservation experience was a harsh lesson for the Forestry Department early in its foray into the creation of Forest Reserves.

The creation of Forest Reserves, such as the Kambui Hills, was ultimately meant to be the cornerstone of colonial forestry in Sierra Leone, providing a legal, physical and philosophical foundation for future forestry and forest conservation initiatives in the country. And, for at least the first few decades of 'scientific forestry' in colonial Sierra Leone (and sub-Saharan Africa in general), forest reservationism was the dominant philosophy of how forest conservation should be achieved. Indeed, it was an objective that was readily quantified. Within colonial forestry literature, it was a regularly stated objective that between 25 per cent and 30 per cent of the land of a tropical country should be put aside for the creation of Forest Reserves.[17] This objective became a litmus test for measuring Sierra Leone's forestry progress, as a 1929 Government dispatch reveals:

> The total reservation today is only 243 square miles, showing an annual reservation of 24 square miles since the formation of the

> [Sierra Leonean Forestry] Department (in 1912), that is to say 0.78% of our territory instead of 25% which I take to be the absolute minimum.[18]

The struggle with the Kambui Hills Forest Reserve was symptomatic of the struggles of creating Forest Reserves in general. Lack of staffing resources and chiefdom resistance made it an extremely complicated process. It had taken six years and numerous direct interventions by the head of the Forestry Department and the Kenema District Commissioner to get the 147 km^2 Kambui Forest Reserve declared, which was equivalent to around 0.21 per cent of Sierra Leone's land area. Even then, its eventual gazettement had in part been achieved through a fortunate outcome with local chiefdom politics (i.e. Paramount Chief Humonya being deposed by her own people). The slow process surrounding Kambui was not unique. The Loma Mountain Forest Reserve in the north of Sierra Leone, for example, was first planned and surveyed in 1918; however, it was not formally gazetted as a Forest Reserve until 1952, suffering from similar gazettement issues as Kambui Hills.

Forestry Laws

The procedure for realising forest reservation in Sierra Leone was underpinned with a Forestry Law that was passed in 1912, formally called *The Forestry Ordinance, 1912*. As noted in the previous chapter, it was formally drafted by Lane-Poole; however, for the most part, it was copied and pasted from the *Southern Nigerian Forestry Ordinance*.[19] The process of establishing Forest Reserves in most of Sierra Leone required that the relevant Paramount Chief (or Chiefs if the proposed area crossed chiefdom boundaries such as with the Kambui Hills) must formally request to have the Forest Reserve created. The indirect rule approach adopted by the British also meant that Forest Reserve land would still be legally owned by the Chiefs; the Forestry Department would in effect be leasing it from chiefdom authorities. It was a weak legal apparatus to establish control over forested areas, and ultimately the Forestry Department's main path to achieving its forest estate was through convincing Paramount Chiefs of the importance of colonial forestry science, as a review of Sierra Leonean forestry in *The Empire Forestry Journal* lamented:

> Ultimately the solution seems to lie in the education of the [Sierra Leonean] native community to the value and importance to them of the preservation of a sufficient number of forest estates to ensure a permanent supply of timber and forest produce for their own and subsequent generations. (So easily written – so hard to achieve!)[20]

Reportedly, never once during the colonial era was there a spontaneous request from a Paramount Chief to have a Forest Reserve created on their land.[21] As such, 'requests' for a Forest Reserve ultimately required political interference, as Sierra Leone's Conservators of Forests explained in 1929:

> The first essential is a 'Request', addressed to Government by the Tribal Authority, for the land to be created as a Forest Reserve. Tribal Authorities do not submit such requests on their own initiative, and it is the work of Political Officers, in suitable cases, to use their influence to procure them.[22]

A proposed solution often put forward by colonial foresters in Sierra Leone to resolve this situation was to make a change to forestry laws and allow for the creation of Forest Reserves through direct injunction by the Forestry Department.[23] The colonial government in Sierra Leone, however, resisted such changes[24] and appeared to be averse to making changes to laws that might cause widespread resentment among chiefdom authorities.[25] Likely, the government was still conscious of the 1890s Indigenous 'Hut Tax' rebellion and therefore was cautious about overreaching politically. The Imperial Forestry Institute summed up the approach in its 1936 Half Year Newsletter: 'It is the policy of [the Sierra Leonean colonial] Government to interfere as little as possible with tribal habits and customs and to let change and progress be gradual.'[26] This meant that the Forestry Department was in a relatively weak position with respect to establishing environmental hegemony across the Protectorate. The tension in dealing with the chiefdom-dominated Protectorate was a key issue for forestry throughout the colonial era in West Africa, causing intense friction between 'scientific' forestry objectives and the political 'realities' of the colonial governance system.

Stagnation

The Forestry Department in its fledgling years was ultimately a precarious agency that was constantly understaffed, poorly funded and had limited administrative power, a situation that was exacerbated during the outbreak of World War I in August 1914. All of Lane-Poole's Assistant Conservators of Forests – except one, Kenneth Burbridge, whom Lane-Poole disliked – joined the army and left their forestry posts. Lane-Poole would also have likely served, but due to him missing his left hand (and having a hook in its place) – the result of an earlier hunting accident – he had to settle for Assistant Censor work in Freetown: monitoring wartime communications for sensitive information. Either way, time, staff and resources were effectively non-existent during the war period.

After Lane-Poole left, he was replaced by a colonial forester named Lionel Palfreman. His name can be found in a 1958 *Empire Forestry Review* article, where he is included in a list of 'legendary … pre-first war era' colonial foresters. Despite this accolade, however, it is difficult to find any in-depth information on Palfreman's time working in West African forestry. Palfreman's four-year tenure as head of Sierra Leone's Forestry is the only period during the entire colonial era that forestry annual reports are unavailable. Nevertheless, it can be pieced together that Palfreman started working as a forestry officer in Nigeria as early as 1905; he held this post until at least 1910 and perhaps until as late as 1917, when he took up his appointment as Sierra Leone's Conservator of Forests. Perhaps due to frustration with the slow pace of forestry in Sierra Leone, he left his position as head of Sierra Leone's Forestry Department in 1921 to become once again a subordinate forester officer in Nigeria. He chose a demotion to escape Sierra Leonean forestry.[27] Palfreman was replaced with another colonial forester who worked in Nigeria, Leonard Acton King-Church. He was born in 1876 and received his early forestry education at Lancing College in the United Kingdom. He only lasted just over a year, leaving in 1922 to take up a post on the Gold Coast (modern day Ghana), where he worked as a Conservator of Forests until his retirement in 1931.[28] Sierra Leone, it would seem, was the least desirable location in West Africa for British foresters. As one Sierra Leone colonial forester lamented:

> The Sierra Leone Forest Service has already suffered severe misfortunes; it was exceedingly hard hit during the [First World] War; when

> it had almost recovered, it was again reduced to impotence by the Lands and Forests Department. Conditions for long continued stability and security are more necessary for success in Forestry than in any other human activity. Except for two short periods immediately before and immediately after the [First World] war, and an even shorter period since last September, the Forest Service has been acutely conscious of the absence of these conditions.[29]

The frustrations of colonial Forestry Staff in Sierra Leone were palpable, with the acting Conservator of Forests, Eric MacDonald, writing a letter to the Colonial Secretary Office declaring that the Department should be closed down due to the pathetic support the colonial government had shown to forestry during its first twenty years in Sierra Leone:

> [The Sierra Leone Forestry Department] had been strangled soon after birth by the lack of continuity in Forestry work caused by outside interference with the Forest Service for the usual reason of temporary expediency. The regrettable but inevitable consequence of this interference is that, with the exception of a number of Protection Forest Reserves, there are now no noteworthy results whatever to show after a Forest Service has ostensibly been in existence here for nearly twenty years ... In view of the fact that this Government is apparantly [sic] unable to guarantee any stability to the Forest Service, and has in the past sacrificed and is apparantly [sic] prepared in future to sacrifice this Service on the altar of temporary expediency, I cannot find any justification for the maintenance of a Forestry Department in Sierra Leone.... In view of the past record, no attempt should be made to carry on the present Department, nor to re-establish it if abolished until ... the Government can accept the establishment of a Forest Service on some footing would make it secure financially and technically.[30]

After much debate,[31] however, in 1930 the Colonial Government decided against Macdonald's recommendation, and 'Forestry in Sierra Leone was once again entrusted to a Forest Department', with the irate Eric Macdonald being appointed to the position of Conservator of Forests.[32] Its budget, however, was minimal, and therefore, due to fiscal constraints, the creation of new Forest Reserves was postponed.[33] The regional forestry offices were closed, and there was a massive retrenchment of staff.[34] An article in *The Empire Forestry Journal* adjudged the situation succinctly: 'The annual report of the Forest Department of

Sierra Leone for 1934 shows the Department to be in a state of suspended animation hardly distinguishable from death.'[35]

Reservation Revival

A small revival in forest reservationism occurred in Sierra Leone during the 1940s and 1950s, namely through the creation of Protected Forests, a newly created type of protected area. The creation of these Protected Forests appears to be a direct reflection of the Forestry Department's previous challenges in regards to reservationism. H. Maughan-Brown, a colonial forester in Sierra Leone during the 1940s, summed up the situation in the popular colonial forestry journal *Farm and Forest*:

> We have ... the problem that while a vast expansion of reservation is essential for a balanced ecology, the people, whose co-operation is necessary, are antagonistic to a reserve as they conceive it. The obvious solution is to present reservation in a palatable form; to get the people to realise that this does not mean the alienation of their land, but rather its employment to preserve as far as possible, conditions favourable to farming while the reserved areas will at the same time yield an economic crop. This is being achieved by working through the Native Administrations and by concentrating at the outset on the 'Sei Bush' type of reservation.[36]

Sei bush (also known as *Salei* bush) is a term used by the Mende in the south-east of Sierra Leone to refer to strips of forested areas that were maintained alongside paths and rivers.[37] These areas of forest were preserved as a part of old native laws, as they offered shading for those travelling along the paths and rivers, as well as suitable barricades for ambushing enemies.[38]

Such an approach to reservation appears to be the innovation of William D. MacGregor, who headed the Forestry Department between 1939 and 1947, although similar approaches were adopted in other parts of the British colonial empire in Africa.[39] The initial design of Protected Forests (also sometimes called Native Administration Forests) was developed in conjunction with Chiefdom Native Administrations (local chiefdom authorities) in 1941, where they voted to revive and protect Sei bush sites along paths and roadsides (200 feet on each side). MacGregor noted that the linear type of forest spread (Sei bushes) had the advantage in that the 'sacrifice' of land would be spread over a larger number of people, each only having to forgo a small portion and thus lessening any resistance to such reservation.[40] The appeal to the Chiefdom Native Authorities to such a form of 'forest reservation' was the ability to maintain direct control over their forest resources – that is, Forest Reserves that had been established in the 1910s and the 1920s in the Protectorate, which, although owned by the Paramount Chiefs, the control, management and use of these spaces was ceded to the Forest Department. Protected forests, on the other hand, were to be completely controlled by the Chiefdom Authorities, with the Forestry Department's role restricted to technical advice and financial support.[41]

For the most part, these Protected Forests were designated as 'production reserves' and were supposed to supply fuelwood and building poles for chiefdom communities,[42] although they were also seen to offer some conservationist functions as well. MacGregor, the head of the Forestry Department during the Protected Forest apogee in the 1940s and 1950s, mapped out an ambitious vision:

> Increases in Native Administration forest areas [Protected Forests] will gradually cover the whole country with a network of forest belts along roads and rivers. In addition to this network they will provide for town supply forests and at least the minor catchments areas. They will have great value in maintaining favourable climatic conditions for agriculture and in providing readily accessible sources of forest produce. Though their aggregate area, expressed as a percentage of the whole country, may not be large, they will be most favourably situated for both protective and productive purposes.[43]

The Forestry Department often provided fast-growing tree species to be planted in such areas.[44] Overall, Protected Forests seemed to present a win-win situation for realising colonial forestry objectives. It allowed the Forestry Department, on paper at least, to expand its classified Forest Estate towards its desired objective of 25–30 per cent, while the chiefdom authorities were given opportunities to create productive forest areas with direct support from the colonial government.[45] Soon after these initial negotiations with chiefdoms, proposals were made by the Forestry Department to make amendments to Forestry Legislation so that Protected Forests could be formally recognised in law.[46]

The creation of Protected Forests proved to be extremely popular, and during a fifteen-year period (1943–1958) over 150 of them were formally gazetted by the Forestry Department. It was a struggle for the Department to keep up with the influx of Protected Forest requests due to the limited amount of forestry staff, and for the first decade, working on a backlog of requests was the norm.[47] At one stage, it was even suggested that Protected Forests might completely supplant Forest Reserves as the main form of protected area in future.[48] Nevertheless, increases in the Forestry Department's prominence in terms of afforestation and exploitation activities apparently provided the necessary impetus of getting Forest Reservation back onto the agenda. As one Conservator of Forests noted:

> For the past decade and more there has been almost a complete cessation of forest reservation – a negative policy for which various reasons was enforced on the department. Certainly there were no spontaneous requests, nor was it considered that requests could be induced by persuasion. There is, however, a very real change taking place. The advent of sawmilling, and the more general awareness of the importance of forestry are having their effect and it is anticipated that requests for reservation will be easier to obtain. But there will be recalcitrant cases to deal with and, to meet these exceptional (it is hoped) cases, compulsory legislation was submitted during the year and is now under consideration.[49]

The compulsory legislation to create Forest Reserves without chiefdom approval never emerged. Nevertheless, the Forestry Department – thanks to extra staffing, the notable successes of exploitation forestry programmes (discussed in the next two chapters) and active support from District Commissioners[50]– did manage to get forest reservation back onto the agenda, with eighteen Forest Reserves established between 1946 and 1961, the last fifteen years of the colonial era. This included the Loma Mountains Forest Reserve, the largest Forest Reserve in Sierra Leone to date. This was no mean feat given the Reserve's remote location and difficult terrains, as one Conservator of Forests recounted: 'The Loma Mountains area includes the highest mountains of Sierra Leone (over 6,000 feet) and is remarkable for the occurrence of *triplochiton*, found nowhere else in the country, but [the] practically gregarious negotiating, demarcating and surveying the boundary of 62 miles was a lengthy and arduous operation.'[51] The forest reservation revival, however, would not last.

Reservation Demise

Like the rest of West Africa,[52] the formal policy approach of 'forest reservationism' for forestry and forest conservation in Sierra Leone came to an end during the early part of the independence era. The main driver, in terms of moving away from forest reservation, was financial. An independent Sierra Leonean Government had considerably fewer resources at its disposal than its colonial predecessor, and in such a tight fiscal environment, forestry budgets were usually one of the first ones to be cut. Its budget had already been substantially reduced just before independence during the mid 1950s,[53] and this trend was to continue into the independence era. A second problem for reservationism was an overall disinterest in forestry in Sierra Leone. During the late 1950s, the Forestry Department had organised 'tours for boys' from Sierra Leone's top secondary schools to visit the Forestry Department operations around the country. The tours were heavily publicised on radio and in newspapers and were aimed at facilitating the 'rapid Africanisation of the Department'[54] as a part of the British colonial government's plan to transition Sierra Leone towards independence. A newspaper article about the visit, which was cut out and saved by the Forestry Department, was hopefully titled 'Boys Given an Idea for a Career',[55] while broadcasts about the visit were also made on the Sierra Leone Broadcasting Service (SLBS) radio news service. The 'tour for the boys' lasted an impressive eight days and involved extensive travel to different forestry sites in the east of Sierra Leone. The Forestry Department, however, struggled to attract local Sierra Leonean candidates:

> The annual visit of Freetown schoolboys to inspect current forestry operations in Bo and Kenema took place in May, 1959. Although many of them expressed surprise and interest in what was shown [to] them, it is unfortunate that there have been no subsequent enquiries regarding vacancies in the Department.[56]

Such a dearth of applications meant that in the 'scramble' to Africanise the Department before independence in 1961, there were times that the Department was forced to accept candidates who had little suitability and/or aptitude for forestry activities. The final problem was that the creation of Forest Reserves in Sierra Leone was never popular among the country's rural inhabitants, and, unlike its colonial predecessor, the newly independent Sierra Leonean Government (during its early years at least) had to rely on popular democratic support to ensure its continual re-election. Given this overall situation, it is therefore unsurprising that the beginning of independence ultimately spelt out the beginning of the end for forest reservationism. Notably, in Sierra Leone, it was not just an end to forest reservationism in terms of political and economic realities but also a philosophical end. As the country moved into independence, the cessation of creating first Protected Forests then Forest Reserves was formalised in policies.

Sierra Leone was granted independence in 1961, and two years later, for the first time, a Sierra Leonean, Joshua S. Sawyerr, was appointed head of the Forestry Department. In a Department that had struggled to attract talented Sierra Leonean recruits, Sawyerr was an exception. He proved to be a rising star, ascending through the Department's different pay-grade positions at a relatively rapid pace. In the late 1940s, he completed a Forest Assistant's course at the Ibadan Forest School in Nigeria and then received a scholarship to study a BSc in Forestry at the University of Edinburgh, which he completed in 1953. After this, he was appointed Assistant Conservator of Forests, and from there it was only a matter of time before he was promoted to the head of the Department.[57] In 1964, Sawyerr authored an extensive report titled *Ten-Year Plan of Forestry Development for Sierra Leone*, which, as the title clearly suggests, provided an aspirational outline for how forestry would operate in Sierra Leone for the next decade.

One of the most notable features in the report was its lack of enthusiasm towards Protected Forests. It noted that foresters needed to 'view more critically proposals from the Tribal Authorities for the creation of Protected Forests'[58] and declared that the previous approach of creating Protected Forests along roadsides should be discouraged; rather, Protected Forests should be in the form of larger contiguous blocks. The report also recommended that some Protected Forests should be 'consolidated into Forest Reserves, whereas others will remain as Protected Forests or be allocated to villages as village [community-run] forests'.[59] Thus, while it was not explicitly stated, the report ultimately insinuated an end to the policy of creating Protected Forests. Indeed, the last ever gazetted Protected Forests had been established four years earlier in 1960.

The move away from Protected Forest creation did not start with Sawyerr; it had been a gradual move by the Forestry Department over the previous decade. While the early success of Protected Forest creation was celebrated by some people in the Forestry Department,[60] others in the Department were not so enthusiastic about the new classified forest type. As the 1951 annual Sierra Leonean forestry report details:

> A large part of the Department's resources has become so deeply involved in a large number of small Protected Forests that more urgent and important work cannot be undertaken. Admirable though they are for local supply purposes, these Protected Forests can never make any substantial contribution to the protective forest estate and a country which has less than 4 per cent of its area under Forest Reserves must give prior attention to the establishment of an adequate area of reservation. Preliminary steps were then to redirect efforts without sacrificing work already done and without neglecting further development of Native Administration forestry.[61]

In the context of the Department's forest estate objective of 30 per cent of Sierra Leone's land area, there was certainly a point to this argument. By 1976, there were 155 formally gazetted Protected Forests in Sierra Leone; however, their average size was only 2.3 km^2 and therefore collectively amounted to only 0.49 per cent of Sierra Leone's total land area. Forest Reserves, in contrast, tended to be much larger (an average of 69 km^2), with the Loma Mountain Forest Reserve by itself almost covering a larger land area than the collective 155 Protected Forests. Furthermore, many conflicts were arising from the creation of Protected Forests, as communities were not receiving adequate financial compensation, and, in some cases, chiefs were using the policy to enact strategic land grabs. Subsequently, it was decided that 'further proposed reservation of this type was [to be] discouraged [from] 1956'.[62] Overall, what this indicates is how forest policy objectives were often entangled closely with the personal beliefs of different heads of the Forest Department, and therefore the rise and fall of programmes such as the Protected Forests were contingent on who was running forestry operations.

Nevertheless, while Sawyerr might have been indifferent towards Protected Forests, his approach to forestry was closely aligned with that of many of his colonial predecessors. His reports espoused the common perception of Sierra Leone having a heavily degraded and rapidly deforesting landscape.[63] Furthermore, he noted the fundamental role that the creation of Forest Reserves had to play in the country's future, offering a slightly more modest forest estate objective:

> The simple thing is to have at least 20% of the area of the country reserved as forest and until this backbone is established the main resources of the Forestry Division will be used unswervingly to this end.[64]

This rhetoric was matched with action, and during Sawyerr's first year as head of the Forestry Department, five new Forest Reserves were proposed.[65] Thus, while Sawyerr, with the newly independent Sierra Leonean Government, had less fiscal and human resources at his disposal in comparison to his colonial predecessors, his vision for forestry in Sierra Leone closely aligned with the colonial vision. Things were to change with his successor.

Musa B.D. Feika formally replaced Sawyerr as head of the Forestry Department in 1975, although it appears that he was informally in charge of things a couple of years before his formal appointment. Feika, it would seem, was more of a politician than a forester. The Feikas were a prominent family in the east of Sierra Leone and had held the chieftaincy of Niawa Chiefdom for many decades.[66] Musa Feika himself was the son of a Paramount Chief, and given this privileged position he was afforded opportunities to study at Sierra Leone's two most prestigious secondary schools, Bo Government Secondary School and the Prince of Wales Secondary School, before completing a Graduate Certificate in Education (GCE) at Sierra Leone's main university, Fourah Bay College, in 1954. He was then awarded a Kenema District Council Scholarship (his father's chiefdom was within the Kenema District) to study a BSc in Forestry at the University of Oxford. Feika, however, proved to be an unexceptional student, and while the Colonial Public Service Commission – the UK-based agency overseeing colonial appointments – wanted him to be slotted into the Forestry Department after graduation, the acting Conservators of Forests at the time were staunchly against this:

> Mr Feika would definitely not be suitable for this [Assistant Conservator of Forests] post as quite apart from his inordinately slow progress his reports have consistently stated no special aptitudes.[67]
>
> As Mr Feika has again failed in two of his degree examinations and cannot now hope to complete his course until March, 1961, at the earliest, there can be no question of holding open the existing Assistant Conservator of Forests vacancy for him until that time. In any case, it is surely obvious that Mr Feika's progress and conduct over the past

> 10 years have been far from satisfactory (compare the case of Mr. Sawyerr, Assistant Conservator of Forests, who had several years of satisfactory experience in this Department before proceeding to [the] United Kingdom where he completed his BSc Forestry course without a single failure) and sufficient to exclude him from any appointment to the Assistant Conservator of Forests grade. Instead, I consider that it would be preferable if he [Feika] were to return to Kenema for appointment as Forester with the District Council.[68]

The Colonial Public Service Commission, nevertheless, continued to push to have Feika assigned to Forestry, noting that although 'his academic progress has been slow … Mr Feika's professor does at least say he is a good practical man'.[69] To the almost certain dismay of the then Chief Conservator of Forest, the colonial secretary, while noting that 'Mr F[eika] unfortunately continues to fail his examinations', he insisted on having him appointed as Assistant Conservator of Forests. There was a dearth of qualified Sierra Leone candidates for the position, and with independence approaching, the Africanisation of departments was of upmost importance.[70] Feika was thus appointed as Assistant Conservator of Forests in the Department in 1961. He repaid the authority's support in him by almost immediately resigning from the position in order to run for a parliamentary seat in the 1962 Sierra Leone general election. He was unsuccessful in his bid and afterwards was given a one-year temporary position in the Forestry Department before returning to an Assistant Conservator of Forests position again in 1963.[71] For the first few years of his position, he was assigned to re-establishing a rubber production industry in Sierra Leone,[72] a task that proved to be unsuccessful.

Feika likely ran as a candidate for the All People's Congress (APC) political party at the 1962 election. While the APC lost this election, it did eventually come to power in 1968 and stayed in power until 1992. This would explain Feika's promotion to head of the Forestry Department in 1975 in spite of a thus far undistinguished forestry career. The mid 1970s was a time when Sierra Leone was going through major political changes, with a move away from democracy towards an authoritarian patrimonial state – epitomised by the creation of a one-party state in 1978.[73] Siaka Stevens, the country's President, had created an informal patronage system that was rigidly organised around the President's Office: power and influence was largely enacted through a system of patrimonial exchanges and favours.[74] In such a governance system, a politically connected individual such as Feika would have undoubtedly been more useful as head of the Forestry Department than a 'highly skilled' forester. Natural resource management during this era was not based on 'sustainable yields and management' but rather on the ability to extract resources profitably with minimal investment from the central government. In line with this necessity, the power of the Forestry Division in this period became highly concentrated in the Chief Conservator of Forests (i.e. Feika), which 'ensured that the Forestry Division literally spoke with a "single voice" in policy formulation and had personalized links to the neopatrimonial state'.[75] The creation of Forest Reserves, which required direct formal government intervention, was not coherent with such a governance approach, and subsequently the system of forest reservation began to unravel.

Early on in his time in charge, Feika probed the role of Protected Forests, questioning whether it was 'a good move to have them constituted' and noting that there was a 'cloud over their future existence'.[76] In fact, Feika actually oversaw the formal de-gazettement of a number of Protected Forests at the behest of some local chiefdom requests.[77] By 'giving back' Protected Forests to local chiefs, he was able to enact a strategy of patronage that realistically represented little cost to the Siaka Stevens government. Feika did not only oversee a formal end to the creation of Protected Forests but also Forest Reserves. Sierra Leone's last ever gazetted Forest Reserve, the Kandesuri Forest Reserve in Sierra Leone's north, was gazetted in 1976, just after Feika had been promoted as head of the Forestry Department. However, the planning and gazettement had started back in 1964 under Sawyerr.

Unlike his predecessors, Feika saw little merit in the forest reservation approach to forestry in Sierra Leone, instead promoting a philosophical approach to forestry more in line with Siaka Stevens' patrimonial state. Feika outline this view in a Sierra Leonean Forestry Report that was published in 1975:

> The strict conservationist view of forestry as formerly held by the Forest Authority of this country was modified to meet present day requirements by placing a more direct emphasis on the business aspect of it. This change of outlook greatly helped to arouse and develop public awareness of the value of the forest and forestry.[78]

Feika's focus was on exploiting Sierra Leone's forests for relatively short-term financial gain, in particular for the centralised Siaka Stevens government. The long-term forestry reservation approach was not conducive to this modus operandi and was subsequently terminated. Thus, the forest reservation approach involving attempts to create a forest estate cover of 30 per cent of the country was abruptly brought to an end.

Forest Reservation Legacies

The story of colonial forest reservation in Sierra Leone, by its own initial lofty objectives, is for the most part a story of abject failure. While the early aspiration of a 30 per cent land area estate was no doubt ambitious, the Forestry Department did not come anywhere close to this objective, crawling to an estate of a few percent over a sixty-year period. The head of Sierra Leone forestry during the late 1950s provides some reflection on the implications of this failure:

> One of my predecessors wrote that a Forest Department without an adequate forest estate has not fulfilled is primary function. With 4.03 per cent of Sierra Leone enclosed in Forest Reserves and Protected Forests, it cannot be said that this Department has yet fulfilled its primary function.[79]

Even the figure of 4.03 percent was overly generous, the result of a miscalculation; a recalculation of the extent of Forest Reserves using the colonial Forestry Department's annual reports shows that the actual area was closer to 3 per cent (see Figure 2.2). Thus, while the British Colonial Government was successful in securing Sierra Leone as a geopolitical territory in the 1890s, it did not translate into *fait accompli* control over the colony's land, resources and people. Natural resource-managing bureaucracies – departments for mining, agriculture and forestry – ultimately had to undertake a process of 'internal territorialisation'.[80] They had to try and assert their dominance over the landscape using relatively weak laws and limited financial support from the colonial government in the face of a somewhat empowered chieftaincy. Forestry was arguably the weakest of these bureaucracies, with agriculture and mining tending to receive much more political and economic support, as they were seen to offer greater colonial profits in the short term – a situation noted by Sierra Leone's acting Governor in 1941,

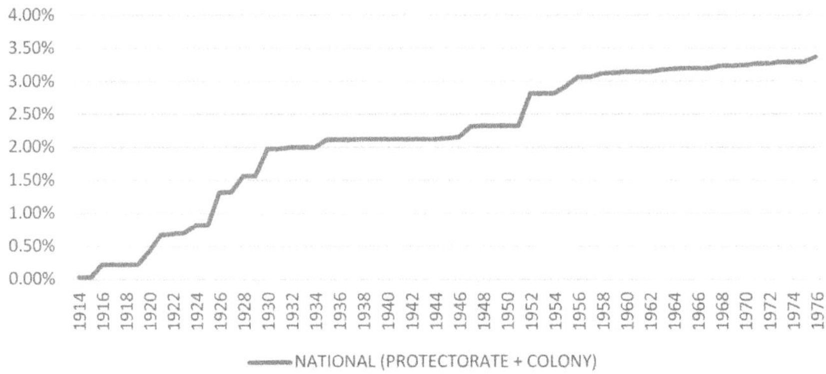

Figure 2.2 The size of the Forest Estate in 'The Colony' and Sierra Leone as a percentage of the total area. Data was collected from reserve gazettement publications in the *Government Gazette* and from Forestry Department reports between 1914 and 1976. Created by the author.

> [Colonial funding] for Agriculture [in Sierra Leone] has increased by 400 per cent in the last twenty years, expenditure on Forestry is at present below the level of 1921, and that there had been an almost complete cessation of forest reservation work over the past decade.[81]

It is instructive that the Forestry Department's greatest reservation success – at least in terms of co-opting the population into its agenda – was through its creation of the classified Forest Reserve type of the Protected Forest. The idea behind this approach to forest reservation did not emerge within forestry schools in Europe but rather came from reading and understanding the socioecological landscape of Sierra Leone. Protected Forests were essentially a revival of Mende Sei bushes, forested areas that were maintained alongside paths and rivers that were preserved as a part of existing chiefdom laws. In contrast to colonial forester Louis Lavauden's declaration that 'the competent man has nothing to learn from the native: on the contrary the native has everything to learn from him',[82] arguably the Forestry Department's most successful period emerged when it tried to 'learn from the native' and emulate existing, non-colonial forest reservation regimes. Nevertheless, this initiative was short-lived, as colonial forestry success was narrowly measured in terms of estate percentages, not by praxis on the ground; and therefore the modest-sized Protected Forests did not sit comfortably with the grander dreams of colonial forestry.

Nevertheless, while the Forest Department only succeeded in securing 4 per cent of Sierra Leone's land area for Forest Reserves, 4 per cent still represents a considerable amount of land that was used by different populations across Sierra Leone. The creation of that 4 per cent had a very real impact on the ground during the colonial era and still has implications in contemporary Sierra Leone. Fundamentally, it is critical to recognise that the creation of Forest Reserves in Sierra Leone (and elsewhere in Africa) was very much a political rather than ecological act. They are, in Vandergeest and Peluso's terminology, 'political forests' – territory that States declare as forest and put under the control of state forestry services.[83] They are the product of social and political as well as ecological processes,[84] and therefore are best understood as socio-natural entities.[85] Forest Reserves, in a sense, were a specific and crucial type of colonial forest practice: they not only fundamentally changed people's lives and livelihoods but created new, almost inescapable means of imagining land, resources and people.[86]

The boundaries of Forest Reserves during the colonial era were ultimately decided in political terms, not ecological ones, and as such had to be 'manufactured' into existence. Their creation was realised through surveys and maps. Therefore, a major component of establishing a Forest Reserve was the creation and maintenance of its boundaries, a demarcated line – physically, legally and philosophically – that was meant to separate society and nature. Those in the highest echelons of the Sierra Leonean Colonial Government had surprisingly great faith in the power of these boundaries and imagined that once enshrined in colonial law the Forest Reserves would also be preserved in permanence on the ground. As one Governor of Sierra Leone recounted:

> The native of the Protectorate is in general law-abiding and ready to respect prohibitions. Political Officers and Forest Officers agree that once a reserve has been demarcated and proclaimed there is little reason to apprehend its infringement by the inhabitants of the surrounding country and that the prevention and detection of forest offence in connection with Protectorate Reserves can be efficiently carried out by much smaller staff.[87]

A combination of this faith in the Forest Reserve and its discrete boundary as a framework for securing forests (i.e. nature), and the widespread perception that any forest under 'native' control was destined to be destroyed, has resulted in the colonial (and now contemporary) forestry discourse that understands Sierra Leone's forests could *only* exist in Forest Reserves. This perception has no doubt been fuelled by maps produced in Freetown during the colonial period, which almost always showed Forest Reserve areas shaded in green. These include the official gazettement maps drawn by forestry staff as well as national maps of the reserves, which were distributed publicly to define the extent of the government's forest estate. As has been widely noted, maps are inherently political,[88] and in the colonial period they were generally produced by powerful political actors and present a colonial view of the landscape.[89] As Braun and Wainwright note:

> Crucially, the 'forest' conservation discourse is not something that existed *independently* from maps, tables, techniques, and practices that made it available to forms of economic and political calculation. One must first see the forest before on can rationalize it.[90]

The historical use of maps and forest conversation can be seen as a form of 'appropriating nature', helping to increase state control over geographical spaces, which are sources of valuable resources.[91] They also, however, produced simplistic understandings of a landscape. In Sierra Leone, the tendency to shade Forest Reserves in green, contrasted against a white background, gives the impression these are the physical 'green areas' (i.e. forests) of the country. These maps, therefore, have helped to legitimise visually the discourse and perceptions of the Forestry Department that forests were 'safe' if they were formally situated within Forest Reserve boundaries. This has helped to reinforce policy approaches such as land acquisition in the form of Forest Reserves to protect the country's forest resources. For example, a Forest Reserve map published in the 1967 book *Sierra Leone in Maps* included text to provide the reader with a narrative to assist with the interpretation of the visual map presented:

> Maps in this volume will confirm the validity of the statement that the present distribution of vegetation in Sierra Leone has been influenced not only by factors like latitude, altitude, relief, soils and rainfall, but also by the work of man [sic]. There is ample evidence to suggest that much of the country was formerly covered in forest, which has been greatly reduced by clearing. Indeed, there is a very little primary forest left, *save in parts of some of the more remote Forest Reserves.*[92]

This discursive transition of the Forest Reserve estate becoming synonymous with forest cover appears to have occurred gradually during the 1930s and 1940s,[93] and by the 1950s, at the Sixth Commonwealth Forestry Conference in 1952, it was formally presented that forests in Sierra Leone only existed inside Forest Reserves.[94] It was a simplification for the calculation of forest cover in Sierra Leone; the percentage of forest cover could be directly equated to the forest estate:

> The remnants of high forest lay in the Western Area hills, along the Liberian border, and in the more sparsely populated parts of the mountainous eastern half of the country. These remnants amounted to no more than four per cent of the area of the country and many of them grow on precipitous hills on infertile soil where their function is more protective than productive. This small high forest estate has already been set aside as Government Forest Reserves.[95]

The creation of 'forests' in the form of Forest Reserves thus was not just a function of the formalisation of the management of forest but also a formalisation of how forests should be understood.

While the maps themselves are a material signification of an appropriated formalised forest, the act of equating a percentage of forest to a forest estate is the most revealing action of all. Forest Reserves were 'political forests'; they existed in the Sierra Leonean legal system and perhaps in the minds of urban-based Government Officials but ultimately had limited relevance or correlation to the material extent of forests in Sierra Leone. Indeed, more recent research in Sierra Leone indicates that there are large extensions of forests outside of Forest Reserves.[96] For example, a recent remote sensing analysis of forest cover change between 2001 and 2010 across the northern half of Sierra Leone showed – in contrast to the popular narrative of rapid forest deforestation in Sierra Leone – that there had been little change in the absolute extent of the region's forest cover: some areas have experienced forest growth; other areas have experienced forest decline.[97] Another study using remote sensing analysis for the same period focused on the northern section of the Kambui Hills Forest Reserve (discussed in the introduction to this chapter) and came to a similar conclusion, noting that:

> The landscape of the area surrounding KHNFR [the Kambui Hills North Forest Reserve] has been highly dynamic over the past decade, with myriad patches of gain and loss in the various vegetation

> categories. The forest reserve itself appears to have remained reasonably intact with only a thin line of forest reduction along its (mainly eastern) fringes, while the entire area surrounding the reserve as a whole appears to have experienced a modest overall increase in forest, or at least dense vegetative, cover. This contrasts strongly with usual environmental narratives in Sierra Leone, which present the country's forest cover as being in a state of constant and/or rapid decline[98]

The study shows why the boundaries of a Forest Reserve are important in framing narratives. The 'thin line of forest reduction along its (mainly eastern) fringes' was mainly due to urban expansion and occurred *within* declared boundaries of the Forest Reserve, while an entire area 'surrounding' (*but not within*) the reserve as a whole appears to have experienced a modest overall increase in forest. Therefore, while it would be accurate to say, from the study's data, that the Kambui Hills North Forest Reserve experienced forest decline *within* its boundaries between 2001 and 2010, it is equally accurate to state that the forest block that the Reserve is part of (and that expands beyond the reserve boundaries) actually experienced an overall increase in forest growth. Seeing forest growth or decline in the Kambui Hills is therefore an ontological question that is heavily premised on the Forest Reserve boundary. As such, Forest Reserves, therefore, should not be seen as a *natural* phenomenon, object or idea but rather as a *social* phenomenon. It is especially political, as what spaces the Forest Department declares to be 'forests' (Forest Reserves) has implications in terms of resources and interventions.[99]

Therefore, a critical question to ask is who has benefited (and currently benefits) from the creation of Forest Reserves in Sierra Leone. Originally in Sierra Leone, Forest Reserves were promoted as being for the use and benefit of the native communities. In part, this was a result of the overall colonial governance system in Sierra Leone. For the Forestry Department, the ability to realise control forests was complicated by the indirect rule approach that was adopted in Sierra Leone, as it gave considerable powers to Paramount Chiefs. This not only meant that the Forestry Department required Paramount Chiefs' consent to establish Forest Reserves but it also meant that the Paramount Chiefs, in the Protectorate, were still owners of the Forest Reserve's land. As one of the historical heads of the Forestry Department noted: 'There [in the Protectorate] land is owned only by native Africans, and cannot legally be alienated to persons who are not native Africans.'[100] Therefore, early on in the colonial era, Forest Reserves in the Protectorate were frequently described as 'communal Forest Reserves', and it was stipulated that they were constituted for the 'use and benefit of the native community which *owns* the forest'.[101]

There appeared to be little that the Forestry Department could do to hasten the expansion of its forest estate into the Protectorate beyond trying to influence and coerce Paramount Chiefs to request the establishment of Forest Reserves. Nevertheless, it appears that Paramount Chief 'control' over Protectorate Forest Reserves was gradually weakened, rhetorically and legally, throughout the colonial era. For example, the emergence of Protected Forests in the 1940s, which were designed to serve Protectorate communities, meant that the status of Protectorate Forest Reserves, originally being for the use and benefit of the native communities, became diluted. Almost as an act of misdirection, by positing Protected Forests as 'Forest Reserves for the community', actual Protectorate Forest Reserves were then posited to be Forest Reserves for the Government. Indicative of this, increasingly the term 'Government Forest Reserve' began to emerge as a common designation during the 1940s and 1950s. Early on in the 1940s, MacGregor, the head of the Forestry Department, noticed this tendency and warned against it:

> In both cases [Protected Forests and Forest Reserves] the land remains the property of the community and the somewhat loosely applied term 'Government Forest Reserve', of Protectorate reserves, is a misnomer and should be dropped.[102]

It appears that little heed was paid to MacGregor's warning, and his successors, who headed the Forestry Department from the late colonial period into the early independence era, frequently utilised the term 'Government Forest Reserve' to help distinguish them clearly from 'Native Administration Protected Forests'.[103] Pauline Von Hellermann has identified something similar in her research in southern Nigeria, where she noted that while native authority forests were designed to be more palatable to chiefdom authorities and appeal to their financial interests, she argues that in practice the move was deceptive and effectively a cover-up to gain territorial control.[104] The mission creep in terms of Forest Reserve control continued into the 1980s. In 1988, a new *Forestry Act* was passed, ostensibly to replace the *Forest Ordinance* of 1912, which had a number of amendments through the years. The new law fails to clearly define who 'owns' Forest Reserves in the Provinces; instead, it rhetorically describes them as 'national forests', implying a sense of underlying state ownership and complete Forestry Department control over these areas.

There are thus two important interlocking threads with colonial forest reservation – that they were seen as the *only* method of protecting Sierra Leone's forests and that the term 'Forest Reserve' became synonymous with 'forests' in the Forest Department and other elite agencies. Thus, creating Forest Reserves became synonymous with forest conservation in Sierra Leone. And second, that the Forest Department, overtime, gradually enacted more legal control over Forest Reserve spaces. There was an overarching colonial logic that presented forest reservation as a solution and thus, as Neumann has suggested, it is more historically accurate to say that Forest Reserves have encroached on peasant communities than the other way around,[105] both in terms of their creation and their legal status. Thus, it is unsurprising that forest reservation has been seen to have 'presented one of the most keenly felt experiences of colonial regimes in Africa and Asia and caused many local protests'.[106] There is a tension of environmentalisms – a colonial conservation ethic looking to control forests for future (national) resource use and protection, and an environmental justice stance for local populations trying to maintain control and rights of access to their surrounding forested spaces.

While the formal process of Forest Reserve creation died out in the 1970s, their creation was fundamental for historical colonial forestry and contemporary Sierra Leonean conservation praxis alike. As the following chapters explore, the sites of Forest Reserves became focal points for colonial programmes surrounding afforestation (discussed in Chapter 3) and exploitation (discussed in Chapter 4). While in the postcolonial era they have formed the basis for wildlife conservation (discussed in Chapter 5). The political basis of Forest Reserves ultimately provided the foundation for future forestry and forest conservation initiatives for the earlier colonial and later Sierra Leonean national government, but at the same time the tension around their control has remained.

Notes

1. A. Sharpe, 'The Hinterland of Liberia', *Geographical Journal* 55(4) (1920), 299.
2. C.E. Lane-Poole, *Report on Forests of Sierra Leone* (Waterlow & Sons, 1911), 9.
3. Lane-Poole, *Report on Forests*, 23.
4. L. Day, *Gender and Power in Sierra Leone: Women Chiefs of the Last Two Centuries* (Palgrave MacMillan, 2012); C. Magbaily Fyle, *Historical*

Dictionary of Sierra Leone (Scarecrow Press, 2006); H.O. Newland, *Sierra Leone: Its People, Products, Secret Society* (Johan Bales and Sons, 1916).

5. C.E. Lane-Poole, *Report on Forest Administration for the Year 1915* (Government Printing Office, 1916).
6. C.E. Lane-Poole, *Report on Forest Administration for the Year 1914* (Government Printing Office, 1915), 6.
7. Lane-Poole, *Report on Forest Administration for the Year 1914.*
8. These conflicts are documented in detail in J. Dargavel, *The Zealous Conservator: The Life of Charles Lane Poole* (University of Western Australia, 2008).
9. A. Abraham, *Mende Government and Politics under Colonial Rule: A Historical Study of Political Change in Sierra Leone 1890–1937* (Sierra Leone University Press, 1978); D. Simpson, 'A Preliminary History of the Kenema Area', *Sierra Leone Studies (new series)* 21 (1967), 52–62.
10. Day, *Gender and Power in Sierra Leone.*
11. C.E. Lane-Poole, Sierra Leone Diary – 4 February 1914: NLA: MS 3799/5.
12. C.E. Lane-Poole, Sierra Leone Diary – 24 November 1915: NLA - MS3799/6.
13. Dargavel, *The Zealous Conservator.*
14. Day, *Gender and Power in Sierra Leone.*
15. Day, *Gender and Power in Sierra Leone.*
16. *The Sierra Leone Royal Gazette* (1920, April 3). Governor's Order: 222–226.
17. L. Lavauden, 'Supplement: The Equatorial Forest of Africa: Its Past, Present and Future', *Journal of the Royal African Society* 36(143) (1937), 15.
18. J.A. Byrne, Government Communicaiton on Re-organisation of the Forestry Department: 15 April 1929: SLNA – A/43/29; see also Colony of Sierra Leone, *Annual Report on the Social and Economic Progress of the People of Sierra Leone, 1937* (His Majesty's Stationary Office, 1938); 'REVIEW: Sierra Leone: Annual Report of the Forest Department for the Year 1930', *The Empire Forestry Journal* 11 (1932), 150–151.
19. Dargavel, *The Zealous Conservator.*
20. J.S. Bates, 'REVIEW: Sierra Leone: Report of The Forest Department for The Year 1936', *The Empire Forestry Journal* 16 (1937), 317.
21. E. MacDonald, Notes on the Committee on land deterioration in Sierra Leone: SLNA – A/42/38; W.A. Fairbairn, 'REVIEW: Report On The Forest Administration of Sierra Leone For The Year 1947', *The Empire Forestry Review* 29 (1950), 82.
22. E. MacDonald, Communication to the Colonial Secretary, 10 April 1929: SLNA – A/43/29.
23. L. Palfreman, Communication from the Conservators Office, Freetown Sierra Leone – 7 March 1918: SLNA – FD 30/1918; E. MacDonald, Government Communication on Re-organisation of the Forestry Department: 10 April 1929: SLNA A/43/29.
24. Commissioner of the Northern Province, Communication to the Colonial Secretary – 5th September 1929: SLNA – A/112/29.

25. W.A. Robertson, Imperial Forestry Institute – Half Yearly news letter (1936): SLNA – A/48/36.
26. Robertson, Imperial Forestry Institute – Half Yearly news letter (1936): SLNA – A/48/36.
27. A.C. Burns, *The Nigeria Handbook: Containing Statistical and General Information Respecting the Colony and Protectorate*, 4th ed. (Lagos Government Printer, 1922); L.H. Dunn, 'Tree-Holes and Mosquito Breeding in West Africa', *Bulletin of Entomological Research* 18 (1927), 139–44; Palfreman, Communication from the Conservators Office, Freetown Sierra Leone – 7 March 1918: SLNA – FD 30/1918; L. Palfreman, Communication to the Colonial Secretary: 17 August 1918: SLNA A/69/30; L. Palfreman, 'The Position of the Crown Colonies – Sierra Leone', paper presented at the British Empire Forestry Conference, London 1920 – Proceedings, Resolutions and Summary of Statements, 1921; L. Palfreman, Christy collection of photographs on Africa: Resthouse and communal plantation, Benin City, Southern Nigeria, 1925: Cambridge University Library – Royal Commonwealth Society Archival Collection –RCS/Y304A; W.M. Robertson, 'L.A. King-Church', *The Empire Forestry Review* 37 (1958), 5; *The Sierra Leone Royal Gazette*, Government Notices 52 (1921), 895; H.N. Thompson, 'The Forests of Southern Nigeria', *Journal of the Royal African Society* 10 (1911), 121–45.
28. J.B. Davy, 'Review: Plants of the Gold Coast by Dr F.R. Irvine', *The Empire Forestry Journal* 10 (1931), 157–58; MacDonald, Government Communication on Re-organisation of the Forestry Department: 10 April 1929: SLNA A/43/29; Robertson, 'L.A. King-Church', 5; *The Sierra Leone Royal Gazette*, Government Notices. *52* (1921): 895; F. Story, 'West Tropical Africa', *The Empire Forestry Journal* 9 (1930), 119–20.
29. MacDonald, Government Communication on Re-organisation of the Forestry Department: 10 April 1929: SLNA A/43/29.
30. MacDonald, Government Communication on Re-organisation of the Forestry Department: 10 April 1929: SLNA A/43/29.
31. J.A. Byrne, Government Communication on Re-organisation the Forestry Department: 29 March 1929: SLNA – A/43/29.
32. E. MacDonald, *Sierra Leone Annual Report of the Forest Department for the Year 1929* (Government Printer, 1930).
33. E. MacDonald, *Sierra Leone Annual Report of the Forest Department for the Year 1934* (Government Printer, 1935).
34. Forestry Department. Question of further retrenchment 1931–1932: SLNA – A/12/31; D.G. Thomas, Reopening Forestry office and stores at Kenema – Communication to the Colonial Secretary – 19 June 1935: SLNA – A/5/45.
35. JRPG, 'Sierra Leone: Annual Report of The Forest Department For 1934', *The Empire Forestry Journal* 14 (1935), 293.
36. H. Maughan-Brown, 'Sei Bush Belt in Sierra Leone', *Farm and Forest* 4(1) (1943), 8.

37. Maughan-Brown, 'Sei Bush Belt', 8–9.
38. J.S. Sawyerr, *Ten-Year Plan of Forestry Development for Sierra Leone: 1964/5 –1973/1974* (Ministry of Natural Resources, 1964).
39. R.P. Neumann, 'Forest Rights, Privileges and Prohibitions: Contextualising State Forestry Policy in Colonial Tanganyika', *Environment and History* 3 (1997), 45–68.
40. W.D. MacGregor, 'Sierra Leone', *News Bulletin of Empire Forest Departments for the First Half of 1941* (1941), 19–20.
41. W.D. MacGregor, *Sierra Leone Annual Report of the Forestry Department for the Year 1943* (Government Printer, 1944).
42. Maughan-Brown, 'Sei Bush Belt', 8–9.
43. W.D. MacGregor, *Sierra Leone Forestry Report for the Year 1941* (Government Printer, 1942).
44. MacGregor, *Sierra Leone Forestry Report for the Year 1941*.
45. MacGregor, *Sierra Leone Annual Report of the Forestry Department for the Year 1943*.
46. W.D. MacGregor, 'Sierra Leone', *News Bulletin of Empire Forest Departments for the Second Half of 1940* (1940), 20.
47. D.H. Hodgson, *Report on the Forest Administration of Sierra Leone for the Year 1948* (Government Printer, 1949); D.H. Hodgson, *Annual Report on the Forest Administration for the Year 1949* (Government Printer, 1951). A.K. Nicol, *Report on the Forest Administration of Sierra Leone for the Year 1955* (Government Printer, 1956).
48. A.K. Nicol, *Annual Report on the Forest Administration of Sierra Leone for the Year 1952* (Government Printer, 1953).
49. MacGregor, *Sierra Leone Annual Report of the Forestry Department for the Year 1943*.
50. D.H. Hodgson, *Annual Report of the Forest Department for the Year 1950* (Government Printer, 1952).
51. R.S. Pelly, *Annual Report on the Forest Administration for the Year 1951* (Government Printer, 1952).
52. For example, see P. von Hellermann, *Things Fall Apart?: The Political Ecology of Forest Governance in Southern Nigeria*, vol. 18 (Berghahn Books, 2013); P. von Hellermann and U. Usuanlele, 'The Owner of the Land: The Benin Obas and Colonial Forest Reservation in the Benin Division', *The Journal of African History* 50(2) (2009), 223–46.
53. Sawyerr, *Ten-Year Plan of Forestry Development*.
54. *Daily Mail*, 'Boys Given an Idea For a Career', May 7, 1959.
55. *Daily Mail*, 'Boys Given an Idea For a Career'.
56. I.G. Bulmer, *Report on the Forest Administration of Sierra Leone 1959/1960* (Government Printer, 1960).
57. Hodgson, *Annual Report on the Forest Administration for the Year 1949*; Hodgson, *Annual Report of the Forest Department for the Year 1950*; A.K. Nicol, *Report on the Forest Administration of Sierra Leone for the Year 1954*

(Government Printer, 1955); R.S. Pelly, 'Sierra Leone', *News Bulletin of Empire Forest Departments for 1950* (1950), 60–61.

58. Sawyerr, *Ten-Year Plan of Forestry Development*, 14.
59. Sawyerr, *Ten-Year Plan of Forestry Development*, 14.
60. D.H. Hodgson, *Report on the Forest Administration of Sierra Leone for the Year 1947* (Government Printer, 1948); Hodgson, *Report on the Forest Administration of Sierra Leone for the Year 1948*; Hodgson, *Annual Report on the Forest Administration for the Year 1949.*
61. Pelly, *Annual Report on the Forest Administration for the Year 1951.*
62. A.F.A. Lamb, *1956 Report on the Forest Administration* (Government Printer, 1975), 4.
63. J.S. Sawyerr, *Report on the Forest Administration of Sierra Leone 1962/1963* (Government Printer, 1963); Sawyerr, *Ten-Year Plan of Forestry Development*; J.S. Sawyerr, *Report on the Forest Administration of Sierra Leone for the Year April, 1963 to March, 1964* (Government Printer, 1965).
64. Sawyerr, *Ten-Year Plan of Forestry Development.*
65. Only one of these, however, was successfully gazetted – the Goloma South Forest Reserve, which was designated in 1965.
66. Day, *Gender and Power in Sierra Leone.*
67. R.M. Palmer, Communication to the Establishment Secretary - 1 May 1960: SLNA - RG 3/2/92.
68. I.G. Bulmer, Communication to Colonial Government Established Secretary - 12 September, 1960: SLNA - RG3/2/92.
69. Public Service Commission. Communicaiton to the Colonial Government Establishment Secretary - 23 May 1960: SLNA - RG3/2/92.
70. Colonial Secretary. Communication on Forestry Department Senior Post Staff Position, 1960: SLNA: RG3/2/92.
71. J.S. Sawyerr, Communication to Colonial Government Established Secretary - 18 July 1963: SLNA - RG3/2/92.
72. Sawyerr, *Report on the Forest Administration of Sierra Leone for the Year April, 1963 to March, 1964.*
73. W. Reno, *Corruption and State Politics in Sierra Leone* (Cambridge University Press, 1995).
74. P. Richards, *Fighting for the Rain Forest: War Youth and Resources in Sierra Leone* (James Currey, 1990).
75. A. Grainger and W. Konteh, 'Autonomy, Ambiguity and Symbolism in African Politics: The Development of Forest Policy in Sierra Leone', *Land Use Policy* 24(1) (2007), 54.
76. M.B. Feika, *Report on the Forest Administration of Sierra Leone 1963/1964 to 1973/74* (Government Printer, 1975), 5.
77. Feika, *Report on the Forest Administration of Sierra Leone 1963/1964 to 1973/74.*
78. Feika, *Report on the Forest Administration of Sierra Leone 1963/1964 to 1973/74*, 5.

79. A.F.A. Lamb, *Report on the Forest Administration of Sierra Leone 1957* (Government Printer 1959).
80. P. Vandergeest, 'Mapping Nature: Territorialization of Forest Rights in Thailand', *Society and Natural Resources* 9 (1996), 159–75; P. Vandergeest and N.L. Peluso, 'Territorialization and State Power in Thailand', *Theory and Society* 24 (1995), 385–426.
81. H.R. Blood, *Reorganisation and Expansion of the Forestry Department* (Freetown: Government Printer 1941).
82. Lavauden, 'Supplement: The Equatorial Forest of Africa', 13.
83. P. Vandergeest and N.L. Peluso, 'Empires of Forestry: Professional Forestry and State Power in Southeast Asia, Part 1', *Environment and History* 21 (2006), 31–64.
84. J. Budds, 'Contested H20: Science, Policy and Politics in Water Resources Management in Chile', *Geoforum* 40 (2009), 418–30; Vandergeest and Peluso, 'Empires of Forestry', 31–64.
85. E. Swyngedouw, 'The City as a Hybrid: On Nature, Society and Cyborg Urbanization', *Capitalism Nature Socialism* 7 (1996), 65–80; S. Whatmore, *Hybrid Geographies: Natures Cultures Spaces* (Sage, 2002).
86. N.L. Peluso and P. Vandergeest, 'Genealogies of the Political Forest and Customary Rights in Indonesia, Malaysia, and Thailand', *The Journal of Asian Studies* 60 (2001), 761–812.
87. J.A. Byrne, Government Communication on Re-organisation the Forestry Department: 15 April 1919: SLNA – A/43/29.
88. M. Monomier, *How to Lie with Maps*, 2nd ed. (University Of Chicago Press, 1996); D. Wood, *The Power of Maps* (Guilford Press, 1992); D. Wood and L. Fels, *Rethinking the Power of Maps* (Guilford Press, 2010).
89. J.B. Harley, 'Deconstructing the Map', *Cartographica: The International Journal for Geographic Information and Geovisualization* 26 (1989), 1–20.
90. B. Braun and J. Wainwright, 'Nature, Post-structuralism, and Politics', in N. Castree and B. Braun (eds), *Social Nature: Theory, Practice and Politics* (Blackwell, 2001), 41–63.
91. L. Harris and H.D. Hazen, 'Power of Maps: (Counter)-mapping for Conservation', *ACME: An International e-Journal for Critical Geographies* 4 (2006), 99–130; N.L. Peluso, 'Whose Woods Are These? Counter-mapping Forest Territories in Kalimantan, Indonesia', *Antipode* 27 (1995), 383–406; P.A. Walker and P.E. Peters, 'Maps, Metaphors, and Meanings: Boundary Struggles and Village Forest Use on Private and State Land in Malawi', *Society & Natural Resources* 14 (2001), 411–24.
92. J. I. Clarke, *Sierra Leone in Maps* (University of London Press, 1969), emphasis added.
93. W.D. MacGregor, 'First Impressions of Forestry in Sierra Leone', *The Nigerian Forester* 1 (1940), 20–21.
94. R.S. Pelly, *Statement by the Forest Authority, Sierra Leone Prepared for the Sixth Commonwealth Forestry Conference, 1952* (Government Printer, 1952).

95. Sawyerr, *Ten-Year Plan of Forestry Development.*
96. P.G. Munro and G. van der Horst, *The Domestic Trade in Timber and Fuelwood Products in Sierra Leone: Current Dynamics and Issues Freetown* (FAO/EU, 2012); G. van der Horst and P. G. Munro, *Land Cover Assessment of the Kambui Hills North Forest Reserve and its Surrounds* (Energy For Opportunity, 2012).
97. Munro and Van der Horst, *The Domestic Trade in Timber and Fuelwood Products.*
98. Van der Horst and Munro, *Land Cover Assessment of the Kambui Hills*, 28–29.
99. P. Robbins, *Political Ecology*, vol. 2. (Wiley-Blackwell, 2012), 123.
100. Robertson, Imperial Forestry Institute - Half Yearly news letter, 1936: SLNA - A/48/36.
101. MacDonald, *Sierra Leone Annual Report of the Forest Department for the Year 1934*, 2, emphasis added.
102. MacGregor, *Sierra Leone Annual Report of the Forestry Department for the Year 1943*, 5.
103. For example, see Bulmer, *Report on the Forest Administration of Sierra Leone 1959/1960*; Sawyerr, *Ten-Year Plan of Forestry Development.*
104. Von Hellermann and Usuanlele, 'The Owner of the Land', 223–46.
105. R.P. Neumann, *Imposing Wilderness: Struggles over Livelihood and Nature Preservation in Africa* (University of California Press 1998), 203.
106. Von Hellermann and Usuanlele, 'The Owner of the Land', 223–46.

CHAPTER 3

Plantations

Sometime in the year 1920, a packet containing seeds of the *Gmelina arborea* tree, known in Sierra Leone by its Burmese name *Yemane*, was shipped from India to the Forestry Department in Sierra Leone.[1] Ostensibly, there was nothing remarkable about this shipment. Botanical experiments were a core enterprise of the British Empire, and seeds were shipped to and from colonial posts all around the world to see how (potentially economically valuable) tree species performed in a range of different geographical contexts. Indeed, the South Asian Yemane seeds arrived into Sierra Leone without fanfare. They were stored at the Forestry Department's office in Kenema for over a decade before anybody bothered to plant them and experiment with them.[2] It was just another 'exotic' tree species to be trialled by the Department. Little did the colonial forestry staff know at the time that these seeds, and more specifically the genealogy of trees that would emerge from them,[3] would have fundamental impacts on the economic and ecological landscape of Sierra Leone in the years to come.

Those shipped Yemane seeds first reached Sierra Leonean soil during the 1930s, when the Forestry Department began experimenting with Taungya Plantations – a mixed farm/forest approach that had been developed by the British in South Asia during the nineteenth century – like Yemane the word *Taungya* ('hill cultivation') and its praxis was an import from Burma.[4] Taungya plantations in Sierra Leone generally involved the inter-planting of 'exotic' and indigenous tree species – early on, the usual mix was 50 per cent exotic with 50 per cent indigenous tree species.[5] The Yemane trees grew the fastest out of all the trialled tree species: Thomas Edwardson, an Assistant Conservator of Forests at the

Department at the time, described its growth as being simply 'unbelievable'.[6] The tree was growing faster than in its native Burma, mainly due to the lack of pests,[7] while its seeds proved to be hardy, keeping their viability much longer than any other species.[8] The tree's rapid germination also caught the attention of the famed colonial forester Professor Stebbing from the University of Edinburgh, during his visit to Sierra Leone in 1934. He later recounted its amazing growth to the head of the Forestry Department in nearby French Côte D'Ivoire.[9] Soon after this, the French Government organised for the tree to be imported from Sierra Leone into different French colonies.[10]

Nevertheless, despite this rapid growth, and critical acclaim from a prominent forester, during the 1930s the Sierra Leone Forestry Department had an ambivalent attitude toward the tree. For example, at the 1935 British Empire Forestry Conference, the Department formally observed:

> *Gumhar*[11] [Yemane] grows exceedingly well and is extensively planted in Taungyas. But it is not seriously regarded as much more than a nurse for more valuable species ... The future of *Gumhar* itself is still uncertain.[12]

Despite their status as a 'nurse species', the Yemane saplings continued to outperform all other planted species in terms of growth and fecundity. This caused problems. In 1940, the Department found that they had a massive excess of Yemane saplings and advertised in *The Sierra Leone Royal Gazette* that they needed to 'dispose' of 17,000 of them among the general public. This compared with only 200 saplings of *Cassia siamea*, the other main 'exotic' tree being planted at the time.[13]

The Forestry Department's attitude toward Yemane gradually shifted from the mid 1940s onwards as the tree's increasing commercial potential was realised (see Figure 3.1). Thinnings from early plantations were converted into roof shingles by the sawmill in Kenema or used to make electricity transmission poles or sold locally for construction purposes.[14] The 1950s mining boom in Sierra Leone also provided a substantial windfall for the Forestry Department, with large amounts of Yemane poles (to be used as mining props) being sold to the iron ore mines in the north of the country and the chrome mine in the east.[15] In 1957, one iron ore mine alone purchased 48,500 Yemane poles.[16] Diverse new uses for Yemane were continually being found: Yemane trees were

Figure 3.1 The Forestry Department's stand at the 1963 Agricultural Show in Kenema. The large white boards reads: 'PITSAWN 14 YEAR OLD YEMANE: The logs are from plantations on Protected Forests managed by the Forestry Department with money supplied by way of the district council'. Photograph supplied by Peter Savill.

linearly planted to demarcate Forest Reserve boundaries; boards were exported to Senegal to be manufactured into matchsticks and boxes[17]; saplings were planted in savanna areas to kill grass cover and provide resistance against forest fires[18]; and Yemane woodwork featured on 'floor strips, furniture and weather boarding' in Freetown houses.[19] This thriving Yemane use was noted by one Dr Henry C. Dawkins, a famous silviculturist at the Commonwealth Forestry Institute at the University of Oxford,[20] during his visit to Sierra Leone in 1965. In his field report from the visit, Dawkins celebrated the 'enormous possibilities for profitable forestry in Sierra Leone [thanks to] the excellent performance of Yemane'[21] and recommended continued research about, and expansion of, Yemane plantations. His belief in the tree and its future in Sierra Leone was strongly emphasised in the peroration of his report: 'Sierra Leone's most important forest assets now are its natural forests and *its Yemane plantations*.'[22] In a couple of decades, Yemane had gone from being a peripheral 'nurse species' to being viewed as an 'absolutely magnificent' tree[23] – the key tree species that the Forestry Department should not only organise its afforestation activities around but its forestry activities in general.[24]

The story of Yemane in Sierra Leone illustrates a couple of entwined threads in Sierra Leone's colonial forest history. The first relates to understandings of Yemane and attempts to cultivate different tree species as an afforestation project. Plantation forestry (also known as silviculture) is broadly defined as being the art and science of controlling the establishment, growth, composition, health and quality of forests and woodlands to meet the diverse needs and values of landowners and society.[25] At first glance, this seems to be a reasonably innocuous, even benign, project of just adding more trees to the landscape. The 'diverse needs and values' that plantation programmes served in Sierra Leone, however, were constrained by a modernist colonial forestry vision and therefore sits within a specific perspective of environmentalism – more focus on sustainable timber yields than on environmental justice. The question of whose needs were being met by the afforestation programmes is a pertinent one. The second thread relates to the materiality of afforestation programmes and its challenges. Yemane did not comport to the Forestry Department's expectations. Its dynamic growth and spread gradually led to Yemane becoming a central focus of the Department in the 1960s. However, as this chapter explores, its centrality in afforestation became a critical issue that the Forestry Department had to deal with.

Taungya

Colonial experiments with tree planting in Sierra Leone pre-dated the arrival of Yemane and were formally initiated in 1896 with the establishment of a botanical station in Freetown. Tree seedlings were sent from Kew Gardens in London,[26] and over the next couple of decades there were a number of small-scale attempts to grow exotic tree species (from Australia, South Asia and Central America) at the botanical station.[27] After its establishment in 1911, the Forestry Department co-opted this existing tree plantation programme and set up nurseries throughout the Protectorate. The afforestation programme was designed to complement the forest reservation programme: through reservationism, the Forestry Department would secure land; through afforestation programmes it would populate this land with desired tree species. The afforestation quest was encapsulated within the colonial vision, shaped by desiccation and climax ecology perspectives that saw Sierra Leone as having a heavily degraded landscape.[28] The aim was to 'repair' the Sierra

Leone landscape by producing more forests, and grandiose visions of 'reclothing' Sierra Leone with trees emerged early on. For example, in 1918 the Colonial Secretary pondered:

> The CoF [Conservator of Forests] wants 30 percent of the area of the [Sierra Leonean] Protectorate reserved and reafforested. The area of the Protectorate is 27000 square miles. Thirty per cent of this is 8100 square miles. The cost of reafforestation at £5 an acre works out at £32000 a square mile. The CoF's estimate of the cost of this item is therefore £25,920,000. Is this figure correct? … I should now like him to give an approximate figure showing what is a fair rent or price per acre to pay to Chiefs for such land … I should like also to know whether the staff he has asked for will suffice to supervise this reafforestation. If not, what staff will be required?[29]

The above ambitious (and expensive) vision was never initiated, and, ultimately, during the 1910s and 1920s few afforestation activities actually occurred in Sierra Leone due to staff shortages, the overt focus on reservation and the general torpor in forestry activities during the departmental amalgamation with agriculture.[30] An afforestation programme was eventually commenced in 1930; this was largely a response to the Colonial Government's ultimatum in 1929, when the Forestry Department was ordered to expand its activities beyond reservation to include more profitable programmes (i.e. plantation work, timber exploitation) or face being decommissioned.[31] The Forestry Department chose the former and started to develop larger-scale experimental plantations.

As noted in the earlier Yemane vignette, the Department's main approach to afforestation was via the *Taungya* method – a mixed farm/forest approach that had been developed by the British in South Asia.[32] Deemed to be a forerunner of agroforestry,[33] it involved land being cleared and planted initially with food crops (usually rice), which was tended by local farmers (see Figure 3.2). Seedlings of a desirable timber species were then planted on the same plot of land, with farmers then tending to the food crops *and* the timber trees. After a few years, once the tree canopy had closed, the land was used solely for producing timber, which was 'felled leaving stumps about 3 feet (1 metre) high. [And] the cut vegetation is allowed to dry and [is] then burnt'.[34] The cycle was then repeated after the timber was harvested.[35] Local farmers were usually able to keep the harvests from the food crops, while

Figure 3.2 Forestry Department Taungya Plantation in Kambui Pass. Photograph taken in 1934, by Stebbing. Lantern slide Va2, in the lantern slides collection at the Radcliffe Library. Used courtesy of the Institute of Archaeology, University of Oxford.

the timber trees produced on the land were owned (and harvested) by the Forestry Department. In Sierra Leone, small financial advances were sometimes given to the farmers as an incentive to engage in the work. However, overall, the approach represented a significant saving in labour for the Forestry Department:

> The advantages of the system are that the heavy cost of clearing the ground is not borne by the Department but by the farmer, and that the trees have the benefit of fairly close cultivation of the soil during the initial critical period of their life – also without cost to the Department.[36]

Plans to introduce Taungya forestry in Sierra Leone started in 1929, with the first experimental plantations established a year later.[37] To begin with, a major nursery was set up in Tabe in 1930, a small Forest Reserve south of Sierra Leone, and for the first few years it was the major supplier of trees for the Taungya programme.[38] Plantation trees were initially raised in the Tabe nursery until they reached a sufficient height; they were then transported in baskets to 'flying nurseries' (small nurseries where plants would be located for a short basis) at each of the Taungya sites, before being planted. The initial focus was on 'exotic' tree species, with fast growing native tree species being co-planted with these (usually a 50/50 mix) to provide shading in order to protect them. Notes taken by Forest Department officials in 1935 provide some insights into the logic of this approach and its challenges:

> The object aimed at in the Taungyas is the establishment of good timber trees. There are many excellent local trees, but I have not found any exotic which appeared to me to be likely to run successfully for a full rotation. On the other hand local tree species regenerate in nature in more or less dense shade and have been found not to thrive when planted in an open rice farm. A nurse was therefore required ... The exotic *Gmelina arborea*, Linn was decided upon. Its early growth is exceedingly rapid and it might possibly run on and produce timber.[39]

The Department, however, in the longer term struggled with the 'many local excellent trees', as its knowledge of the silviculture potential of indigenous trees was limited, with only two indigenous tree species – *Terminalia ivorensis* and *Nauclea diderrichii* – ever being successfully planted on a large scale.[40] And, given the lack of funding for in-country silviculture research in Sierra Leone, the Forestry Department instead looked to other parts of the empire for fast- growing species that could be successfully cultivated elsewhere. Taungya was conducted on an experimental scale at a few selected sites around the country throughout the 1930s. They were mixed plots of native and 'exotic' trees, many including Yemane. To facilitate the dissemination of the necessary trees, more nurseries were set up around the country, with young saplings being raised and cultivated until they were ready for plantation work.[41]

This Taungya initiative was widely credited to be a success by local colonial officials, and detailed descriptions celebrating Sierra Leone's Taungya plantations appeared in the seminal text *The Forest of West Africa and the Sahara*, authored by Professor Stebbing;[42] it was described as a 'bold project [that] will be interesting to watch'.[43] The only thing holding back the expansion of the Taungya programme at its early stage was a lack of forestry staff and funds.[44] This issue was to be solved in part in 1938, when the Colonial Government agreed to fund the expansions of the 'popular and successful' Taungya programme.[45] The Taungya plantation programmes received a further boost in the 1940s, with the creation of the classified forest type 'Protected Forests' (as discussed in the previous chapter). These 'Chiefdom Forest Reserves' often covered areas that the Department classified as 'farm bush' and thus were seen as ideal locations for plantation activities.[46] They became the main sites for establishing Taungya plantations until Sierra Leone's independence in the 1960s.

While Taungya was celebrated as a successful enterprise by colonial foresters, this view of success was entrenched in the notion of trying to achieve 'colonial forestry' and not based on the desires of the local population. It was largely conducted in Protected Forest areas, which were designed, in one colonial forestry official's words, 'to present reservation in a palatable form; to get the people to realise that this does not mean the alienation of their land'.[47] But, in practice, it did mean the alienation of their land. The creation of Protected Forests was implemented by taking away small pieces of land from a large number of farmers, without any compensation being paid.[48] These displaced farmers were subsequently 'allowed' to come back and work on the land in Taungya plantations; however, they did so in an indentured form. They were no longer small-holder farmers managing *their* land; instead, access to land was granted by providing their labour to the colonial government. This meant that there was always a ready supply of labour. Unsurprisingly, the Taungya schemes were most successful in areas where available land was scarce,[49] with many farmers having little choice but to work the plantations to realise their livelihoods. They were given no ownership rights in relation to the Taungya land and were usually only allowed to farm for a maximum of two years before being forced to 'move on'.[50] They had no rights to royalties extracted from the planted trees, nor did they have any say in what types of trees should be planted.

Protected Forests and their Taungya plantations were supposed to create local supply forests – providing fuelwood and poles for nearby communities.[51] This seemingly benign objective, however, was never fully realised. To start with, very few of the tree species proved to be useful for poles or firewood.[52] Moreover, when it came to exploiting the timber in Protected Forests, it was harvested by the Forestry Department and sold to commercial interests, notably mining operations. It was only after these commercial sales were completed that, as one forest report in the late 1940s confessed, 'the balance [was] given to local people for their own use'.[53] The community-owned 'Protected Forest' classification appears to have operated as a subterfuge, allowing for gradual dispossession, whereby the Forestry Department was able to take over land and integrate local communities into its broader colonial forestry objectives. Instead of allowing for the creation of local 'supply forests', the Forestry Department had instead maintained control of its plantation forests and in turn produced a modest revenue for its operations.

Taungya forestry, in Sierra Leone and elsewhere, therefore was fundamentally an exploitative practice that was geared towards achieving colonial forest goals.[54] Raymond Bryant, in his examination of Taungya forestry in Burma, comes to a similar conclusion, noting that the approach was not based on premeditated 'scientific' design but was a product of 'acquisitive colonial power and a threatened indigenous people'.[55] The system offered the colonial government a cheap form of labour and the means of controlling an 'otherwise antagonistic population', while for the local population it 'offered a means of protecting something of their lifestyle'.[56] Its success, therefore, was not so much due to its forestry outcomes but rather an approach to co-opt the local population, who, by and large, had been antagonistic towards colonial forms of forestry, such as the creation of Forest Reserves. Taungya was essentially a means for mobilising resident labour to plant economically valuable tree species.[57] The Forestry Department had succeeded in exploiting local labour in order to help create forests amenable to exploitation by colonial-style commercial forestry.

Afforestation programmes in Sierra Leone expanded dramatically from the 1940s until independence. In addition to Taungya, an afforestation programme of 'natural regeneration' was initiated in the mid 1940s and was implemented in conjunction with the Forestry Department's timber-logging operations. 'Natural regeneration' is perhaps misleading, as it involved eliminating any 'useless trees' (i.e. non-economic trees) through a number of anthropogenic activities: brushing (i.e. cleaning) the regeneration area by using basic tools; girdling (suffocating a tree by cutting a ring around its bark); and, perhaps most dramatically, the use of poisons (e.g. sodium arsenate).[58] These 'natural regeneration' programmes were implemented in 'Forest Reserves' (which were the focus of the Department's logging operations, as discussed in the following chapter), while Taungya was used in the small-sized 'Protected Forests'. All in all, the Forest Department's reservation programme had provided the geographical foundation for its afforestation activities.

With Yemane's rapid growth and its increasingly realised commercial applications, along with the 'success' of the Taungya approach – which offered the Department a cheap source of indentured labour – and the increased use of 'natural regeneration' techniques, it is unsurprising that the Department's plantation work was considered one of its greatest achievements during the late colonial era in Sierra Leone. What could go wrong?

Yemane Crisis

The Sierra Leonean Forestry Department, conscious of the dangers of being overly reliant on a single tree species (i.e. Yemane) for it plantation operations, as it could be wiped out by an arboreal disease or be undermined by market fluctuations,[59] experimented with many different tree species to try and diversify its afforestation operations. While the earliest exotic trees that were part of Forestry Department plantation trials were imported from South Asia (e.g. *Cassia siamea*, *Gmelina arborea* and *Tectona grandis*), during the 1940s and 1950s, exotic species were trialled from all parts of the empire.[60] There was no systematic approach to the selection, and it was often based upon the whims of the head of the Forestry Department. For example, when Alan Lamb became Chief Conservator of Forests in Sierra Leone in 1957, he brought with him numerous seeds from his previous postings in British Honduras (1943–1950) and Trinidad (1950–1956).[61] On his departure from Sierra Leone in 1960, Lamb was credited as having done 'much to stimulate [the Forestry Department's] interest in silviculture'.[62] Thus, seemingly capricious decisions about tree species experiments by Forestry Department personnel could potentially have played a significant role in shaping the country's future ecology.

These 'exotic' tree species were used in Taungya plantations and in various attempts to initiate urban forestry in Freetown. This included planting the Cajeput Tree (*Melaleuca leucodendron*) in the swamps in western Freetown in an attempt to help reclaim land.[63] There were also repeated attempts at planting Beaf Wood Trees (*Casuarina equisitifolia*) along beaches in western Freetown to provide shade for beach goers – the attempt in 1948 failed because crabs kept on damaging the trees;[64] the 1959 attempt failed because somebody kept on stealing the saplings; while the 1964 attempt failed due to a fire sweeping through the area.[65] The most novel project was the introduction of Mexican White Cedar (*Cupressus lusitanica*). Grown in a village just outside of Freetown, the tree was sold to (presumably expatriate) residents in Freetown as a Christmas tree. It turned out to be a moderately successful venture, with an 'ever increasing demand' and provided a small source of revenue for the Department during the 1950s and 1960s.[66] Nevertheless, in spite of all of the above plantation experiments, by the early 1960s Yemane had proven to be the only successful large-scale plantation species.[67] This was a problem, as during the 1950s and 1960s the Forestry Department's fears about being over-reliant on a single species began to be realised when widespread dieback was witnessed across many Yemane plantations.

Dieback is a condition in which a tree begins to die from the top of its leaves (crown dieback) or its roots (root dieback). This generally occurs due to disease or unfavourable climatic conditions.[68] Its sudden onset in Sierra Leone, across Yemane plantations that were 15 to 20 years old, represented a key problem of afforestation programmes with exotic species – that the 'success' of an exotic tree cannot be fully evaluated until it has gone through a full growth cycle.

The age of Yemane dieback was particularly frustrating for the Department, as the best time for harvesting Yemane timber was at twenty years, just on the cusp of the common age for dieback.[69] The first case of dieback in Yemane was observed as early as 1947 in some of the older plantations that had been planted at the beginning of the 1930s. The Forestry Department immediately enlisted the help of the Agricultural Department's Plant Pathologist, yet he was unable to come to any definitive conclusion about its cause.[70] Over the next couple of decades, the phenomenon spread, but it proved to be erratic: some trees died of crown dieback and some of root dieback, and some plantations were heavily affected (around 40%) while others were not.[71] Some trees died quickly while others within a couple of metres of them remained healthy and completely unaffected.[72] Theories of what caused the dieback also varied over time. During the late 1950s, it was thought that a root fungus or a pathogen was the likely cause;[73] however, a specialist team from the Commonwealth Forestry Institute, sent to Sierra Leone in the 1960s, concluded that 'site limitation' and 'root competition' was the *most likely* cause.[74]

It appears that there has never been a definitive conclusion about what caused the dieback;[75] nevertheless, it was noted that it was more common in under-thinned plantations and less common in Yemane outside of plantations (e.g. those planted along the boundaries of Forest Reserves).[76] Thus, the prescription for addressing the dieback crisis was more vigorous thinning of Yemane plantations.[77] This was no simple task. Yemane's ability for regrowth was exuberant, and the Forestry Department struggled to contain young Yemane saplings.[78] As one Chief Conservator of Forests complained:

> Trouble has again been experienced in killing coppice growth of *Gmelina arborea* and continued cutting, breakage and painting with butyl ester arboricide have all been tried without success.[79]

Yemane was too successful in its growth and defied management schemes. By the 1960s, the earlier elation around Yemane had somewhat dissipated, and although it remained a key part of the Forestry Department afforestation programme, its future as a profitable programme was increasingly brought into question.[80] The programme, nevertheless, continued with some vigour into the early postcolonial era, even when government revenue for forestry dried up. This was largely thanks to Joshua Sawyerr, the head of the Department between 1963 and 1975, who was a keen plantation advocate.[81] By the early 1970s, however, when Sawyerr's tenure was fading (and his successor's – Feika – was rising), it appeared that the afforestation programme was going to meet the same fate as the reservation policy, which was ceased in 1976.[82] The programme, however, was not only saved thanks to the 'fuelwood crisis' of the late 1970s but it was dramatically expanded.

Fuelwood Crisis

While the Forestry Department's afforestation programme had an economic objective – that is, populate the landscape with commercially exportable trees – it also had a secondary, environmental one. As discussed in the earlier chapters, there was a common colonial perception that Sierra Leone had a heavily degraded landscape and that it was in immediate danger of being further destroyed due to the agricultural activities of the local population. Professor Stebbing, the colonial forester who visited Sierra Leone in the 1930s, popularised this degradation into a metanarrative, arguing that one day soon the Saharan Desert, which lay several hundred kilometres north of Sierra Leone, would consume the entire region of West Africa. He was the first person to popularise the notion of 'desertification':

> And the desert is advancing ... the end is obvious: total annihilation of vegetation and the disappearance of man [sic] and beast from the overwhelmed locality ... It should not be beyond the power of man [sic] to put up a barrier to this threat.[83]

Stebbing saw Sierra Leone as being one of the most alarming cases of degradation and desertification: 'I did not meet any other colony in which "bush" had reached so degraded a stage over so large areas as I saw during an all too short visit to Sierra Leone.'[84] Sir Arthur Hill, the Director of the Royal Gardens at Kew, in response to Stebbing's observation, suggested that nearly a million trees would need to be planted in West Africa to halt the approach of the desert.[85] Initially, Stebbing argued that an extensive Forest Reserve network should be the main barrier[86] but then later agreed that silviculture also had a major role in improving Forest Reserves.[87] Indeed, he used the example of Taungya and the planting of Yemane in Sierra Leone as the right kind of example for addressing the issues of the supposedly approaching desert.[88] The Sierra Leone Forestry Department used this crisis narrative to justify their afforestation focus on exotic rather than native species. For example, at the 4th British Empire Forestry Conference, held in South Africa in 1935, a resolution (No. 7) was passed stating that 'suitable areas of indigenous forest [were] to be preserved in their primeval form'. All those in attendance – the UK government and all of its colonies – fully supported the resolution, with the exception of the delegation from Sierra Leone, who stated that 'such reservations are a luxury which only wealthier countries can afford!'[89] Profitable timber plantations might have been the core objective of the plantation programme, but the environmental narrative of repairing a 'degraded' landscape played an important auxiliary role. This was to change in the postcolonial era, where the environment narrative would become the main driving concern for Sierra Leone's afforestation programme.

The global 'fuelwood crisis' emerged during the 1970s when many studies estimated that Africa's forests would be decimated in the near future due to the fuel requirements of the poor rural population.[90] Fuelwood consumption was rapidly framed as a critical issue of domestic, regional and international environmental concern.[91] At the time, rural (and urban) populations in Africa largely relied on fuelwood (firewood and charcoal), and thus their continuing consumption of this biomass in conjunction with the continent's growing population was seen as a recipe for a Malthusian-style catastrophe.[92] For example, Erik Eckholm, a prominent environmentalist in the 1970s, described fuelwood harvesting as causing nothing short of suicidal deforestation in Africa.[93] The assumptions of a looming crisis subsequently became grounded in studies of the 'fuelwood issue', with reports on Sierra Leone during the early 1980s predicting a local crisis if major policy measures were not adopted:

> Sierra Leone is suffering from … a fuelwood crisis … Some 70–80 percent of the national energy balance consists of fuelwood, which is rapidly being depleted, even along the traditionally forested area on the coast. Fuelwood consumption is approximated at 2.5 million m^3 annually, with Forest Reserves estimated at less than 6 million m^3. Simple mathematics would lead one to conclude that in 3 years there will be no forests left.[94]
>
> Sierra Leone … is experiencing serious energy problems caused by the rate of consumption of fuelwood (firewood and charcoal) … The depletion of Forest Reserves is now a cause for anxiety since the present rate of consumption of fuelwood far exceeds the rate of replenishment of this natural resource, and all indications point to increasing exploitation of this fuel source in terms of the numbers of users and the rate of consumption. This poses a serious challenge for proper forest management and conservation strategies which will ensure adequate supplies of fuelwood while at the same time ensuring a brake on the indiscriminate use of this fuel source.[95]

A major impact of the 'fuelwood crisis' was therefore an adjustment of the old colonial deforestation narratives. During the colonial era, 'local agricultural practices' (swidden agriculture) were identified as the major cause for deforestation. Since the 'fuelwood crisis', however, local firewood harvesting and charcoal making have been added to this narrative. Sierra Leone's forests were now perceived as disappearing due to local agricultural methods *and* local community energy consumption: 'Logging, slash-and-burn agriculture, and the cutting of trees for use as fuelwood are the primary causes of deforestation in [Sierra Leone].'[96] Therefore, even greater blame was placed on local communities for the perceived destruction of the Sierra Leonean forest environment.

Despite repeated predictions, however, the apparently imminent deforestation disaster never materialised in Sierra Leone, and the notion that fuelwood harvesting causes mass deforestation has been thoroughly discredited by numerous subsequent studies.[97] A key observation has been that in most countries (in Sierra Leone and elsewhere) firewood and even much charcoal production depends on fallen dead branches or the residues of farm clearance in shifting agricultural systems and hence involves very little additional tree-felling. Moreover, like most studies of the period, the above quoted reports failed to appreciate that biomass is a renewable resource – that trees grow back – particularly in the fallow systems that dominate tropical agricultural landscapes.[98] Nevertheless, despite the lack of an empirical basis for its claims, in Sierra Leone as elsewhere across the 'developing world' the 'fuelwood crisis' discourse remains influential, and firewood consumption is still demonised as a major cause of deforestation.[99]

The main response to this perceived crisis has been to increase tree-planting programmes in Sierra Leone. Prior to the 1970s, tree planting had mainly been the preserve of the Forestry Department. Post-fuelwood crisis, tree planting became a mainstream development activity. Major international development NGOs, such as CARE International[100] and Plan International; multilateral agencies such as the World Bank, the United Nations Development Programme (UNDP) and the United Nations' Food and Agricultural Organisation (FAO); bilateral donors such as German Technical Cooperation (GTZ) and the United States of America embassy all became involved in financing and implementing tree-planting projects during the 1980s and 1990s.[101] Symptomatic of the importance of tree planting, in 1985 the World Environment Day (celebrated internationally on the 5th June) was officially rebranded as 'National Tree Planting Day' in Sierra Leone.[102] Thus, from the mid 1970s to the mid 1980s, tree planting shifted from being a response to fuelwood issues to being a core project of the Sierra Leonean environmental movement. The act of planting a tree had become a powerful environmental symbol in Sierra Leone.

A number of local environmental NGOs also emerged in Sierra Leone during the 1980s and 1990s (e.g. the Conservation Society of Sierra Leone (CSSL) and the Environmental Foundation for Africa (EFA)) and likewise have adopted tree planting as a key component of their programmes. Most notably, the NGO Green Scenery has tree planting as the core approach of its environmental activities. In 2004, these local environmental NGOs (along with others) formed a coalition known as the 'Environmental Forum for Action' (ENFORAC), and one of their first activities as a coalition was a UNDP-funded tree-planting project around the Freetown Peninsula.[103] The World Bank and the FAO even worked on reviving Taungya forestry during the 1980s but with a controversial difference: farmers would not receive payments as incentives; instead, they would have to pay rent to the Forestry Department.[104] The programme appears to have had very little success, and while some Taungya forestry occurred during the 1980s and 1990s, it was done on a very limited scale.[105] In areas where it did still occur in the 1980s, it had essentially been co-opted under 'agro-forestry',[106] which has a similar approach to farming/forestry integration but under a name with less colonial baggage.

Although much of this tree planting has been occurring outside of its jurisdiction (territorially and administratively), the Forestry Department has nevertheless maintained a central role in Sierra Leone's broader tree-planting proliferation. Perhaps most notably, it has provided 'received wisdom' from its previous plantation experiences,[107] while its nurseries supply many of the trees that are being used in the planting programmes. As such, these programmes have, by and large, continued to use the same exotic trees that were used during the colonial era. Yemane, in particular, is still being widely planted throughout the country, including by the Vice President, as described in the introduction of this book. Indeed, the 5th of June – National Tree Planting Day – is ultimately a mass tree-planting initiative where thousands of exotics trees are planted around the country by the Forestry Department, NGOs, school children, villagers and others. Contemporary tree planting continues to be conducted with an almost religious fervour in Sierra Leone.

Afforestation Legacies

The story of National Tree Planting Day in Sierra Leone is thus a story of material and discursive legacies from the Forestry Department's colonial afforestation programme. The Forestry Department wanted a forest that was easier to count, control, evaluate and utilise.[108] With the use of exotic trees, such as *Gmelina arborea* from South Asia, afforestation initiatives (overtly or inadvertently) appeared to be focused on creating simplified forests: forests that can be easily exploited for timber or managed for ecosystem services. This focus resulted in the creation of an ecological 'tree space' that differed from (pre-colonial) native forests. This in turn helped to promulgate a simplified approach to environmentalism – the idea that afforestation is needed to *repair* a degraded Sierra Leonean landscape and that the planting of trees should always be seen as an environmental beneficial practice. The health of complex forest ecosystems is reduced to a simple enumeration of trees, regardless of their species. As such, many environmental actors in Sierra Leone espouse a contradictory approach – discursively they emphasise the need to protect Sierra Leone's biodiverse primary rainforest, while on a material level their tree-planting actions are producing a very different forest future.

Indeed, given that Yemane has gone from a packet of seeds in 1920 to a tree that can be found all across the Sierra Leone landscape in less than 100 years, it is reasonable to expect that the tree species will become even more prominent in the future. In the last 10 years, Acacias and Eucalyptus trees have also been planted in great quantities in Sierra Leone with unknown long-term environment impacts.[109] In Sierra Leone, there are currently no studies being conducted to understand the ecological impacts that these tree species have on the landscape, nor are there any programmes to monitor their spread. As has been noted, governments in developing countries often lack the resources to be able to track and manage introduced tree species.[110] Indeed, in its biodiversity report in 2003, the Sierra Leonean Government acknowledged that there had been 'too much emphasis on alien or exotic species in afforestation programmes' and that there was a need to 'develop and promote programmes for monitoring, control and surveillance of alien species'.[111] This, however, has not been operationalised in government policy or programmes, and in its 2008 biodiversity report, the Government of Sierra Leone acknowledged that nothing had been done in Sierra Leone to address or even understand if it was an issue.[112] This is a concern, as Yemane, from reported experiences in Sierra Leone, as well as through laboratory testing, has shown to be a tree species that has a tendency to suppress other (indigenous) tree species,[113] meaning that it could potentially be classified as an invasive species in Sierra Leone. Exotic tree species, nevertheless, are still being planted en masse.

Reginald Cline-Cole has suggested that this widespread use of exotic trees, often found in monoculture plantations, is a self perpetuating approach to afforestation in West Africa. This largely links back to the early colonial problems of a lack of knowledge of indigenous tree species and therefore an ultimate reliance on exotics.[114] Subsequently, by planting exotic trees and examining their progress in Sierra Leone, more and more silvicultural knowledge is being produced about these trees, making them even more desirable as a plantation option. Kojo Amanor, with his research in Ghana, has further suggested that they are also popular for contemporary tree-planting projects, as they are easily distinguished from existing trees and can therefore be quantified in monitoring and evaluation programmes,[115] the ability to quantify outputs generally being a key aspect of most donor-funded programmes in 'developing world' contexts.[116] The prevalence of exotic trees has shaped how 'good' environmental practices are realised by organisations in Sierra Leone. A critical question to ask, then, is who benefits from these environmental practices.

During the colonial era, Yemane's success was celebrated when it could provide commercial profits for the Forestry Department rather than for its functional use in the day-to-day lives of Sierra Leoneans. Thus, approaches that aided in the commodification of Sierra Leone's forests were a common theme of plantation programmes; it was, as one colonial forester in Sierra Leone summarised in 1937, about 'gradually converting an *unsightly jungle* into an orderly station which would be of practical scientific and *economic value*'.[117] Plantation programmes had a key objective: to transform Sierra Leone's forests into a more profitable commodity. As one Chief Conservator of Forests plainly noted, 'the objective was to concentrate valuable trees and multiply the output of the forest many fold'.[118] The Forestry Department wanted a forest that was easier for foresters to count, manipulate, measure, assess and exploit[119]– that is, a forest that equated to around half a dozen indigenous tree species and several exotic trees that could be readily exploited for the European markets. As such, the colonial silviculture programmes in Sierra Leone can be understood as highly political – designed to alter the state of the country's forests physically and epistemologically so that they could be situated neatly within the broader colonial capitalist project.

This does not mean that the afforestation programme eschewed environmental concerns, and, indeed, on the back of the influential Stebbing's declaration of West African desertification, it was framed as a necessarily saviour for the region. Yet these environmental concerns were aligned to European and colonial perceptions of environmental issues – they had a consequentialist framing, addressing a (supposed) impending threat, and paid little heed to the procedural concerns. The Taungya plantation was emblematic of this; it was a programme that relied on alienating local populations from the land and then allowing displaced farmers to return under an indentured work programme. Taungya was effectively a means for mobilising resident labour to plant economically valuable tree species.[120] Thus, while the Taungya plantation programme can be celebrated as a colonial environmentalist programme to reclothe the Sierra Leonean landscape, for many local communities it was a story of environmental injustice. Local populations were being co-opted in an endeavour to realise a European colonial environmental objective.

Yemane and its relationship to Taungya, however, did not comport to the Forestry Department's expectations. It went from a 'nurse species' (1930s) to a lauded success story (1940s to 1950s) to a problematic dieback conundrum (1960s). Today, it is no longer a plantation species that is harvested for large-scale timber operations, as envisioned and promoted by the Forestry Department, rather it is a tree that has spread around Sierra Leone (with the aid of tree- planting initiatives) and is being used for a whole variety of needs, including medicines, housing construction and boat making, among other uses.[121] Indeed, recent research has observed that it is the most commonly harvested tree within villages around Sierra Leone, providing important income for many through informal timber trade networks that operate outside of the large-scale Forestry exploitation programme that was formed during the 1940s, as the following chapter explores.

Notes

1. J.E.D. Fox, 'The Growth of *Gmelina Arborea* Roxb. (Yemane) in Sierra Leone', *The Commonwealth Forestry Review* 46 (1967) 138–44; E. MacDonald, *Sierra Leone Annual Report of the Forest Department for the Year 1930* (Freetown: Government Printer, 1931); J.S. Sawyerr, 'National Report on Man-made Forests (Sierra Leone)', paper presented at the World Symposium on Man-Made Forests and their Industrial Importance and Eucalyptus Study Tour, Australia, 1967; E.P. Stebbing, 'REVIEW: Calopepla leayana (LATR.) and the Possibilities of Control. By P.F. Garthwaite (Indian Forest Record (Entomology), Vol. V, No. 2. New Delhi: Government of India Press, 1939. 2S. 6d.)', *The Empire Forestry Journal* 18 (1939), 320–24.
2. MacDonald, *Sierra Leone Annual Report of the Forest Department for the Year 1930*; Stebbing, 'REVIEW: Calopepla leayana (LATR.)', 320–24.
3. Fox, 'The Growth of *Gmelina Arborea*', 138–44; Sawyerr, 'National Report on Man-made Forests (Sierra Leone)'.
4. R.L. Bryant, 'The Rise and Fall of Taungya Forestry: Social Forestry in Defence of the Empire', *Ecologist* 24(1) (1994), 21–26; N. Menzies, 'Three Hundred

Years of Taungya: A Sustainable System of Forestry in South China', *Human Ecology* 16(4) (1988), 361–76.

5. E. MacDonald, Communication to C.V. Wallace: Forestry Department handing over notes, 1929: SLNA - CSO 2/8; MacDonald, *Sierra Leone Annual Report of the Forest Department for the Year 1930*; D.G. Thomas, *Sierra Leone Annual report of the Forest Department for the Year 1935* (Government Printer, 1936).
6. T.E. Edwardson, Sierra Leone Diaries - July 4, 1935: RHA - Mss afr r 210(1).
7. H.R. Blanford, 'REVIEW: Sierra Leone: Annual Report of The Forestry Department for the Year 1938 (Memeographed.) (Government Printer, 1939.)', *The Empire Forestry Journal* 19 (1940), 160–61.
8. Sawyerr, 'National Report on Man-made Forests (Sierra Leone)'.
9. Stebbing, 'REVIEW: Calopepla leayana (LATR.)', 320–24.
10. Stebbing, 'REVIEW: Calopepla leayana (LATR.)', 320–24.
11. *Gumhar* is the Hindi name for *Gmelina arborea*.
12. Sierra Leone Forest Department (1935). Exotics - Statement Submitted by the Forest Department, Sierra Leone, West Africa, in S.L. Forest Department (ed.)., *The Fourth British Empire Forestry Conference*.
13. R.S. Pelly, 'Forest Department - Disposal of Plants', *The Sierra Leone Royal Gazette*, May 14, 1940, 280.
14. D.H. Hodgson, *Report on the Forest Administration of Sierra Leone for the year 1947* (Government Printer, 1948); D.H. Hodgson, *Annual Report on the Forest Administration for the Year 1949* (Government Printer, 1951); W.D. MacGregor, *Report on the Forest Administration of Sierra Leone for the Year 1945* (Government Printer, 1946); I.G. Bulmer, *Report on the Forest Administration of Sierra Leone 1959/1960* (Government Printer, 1960); A.F.A. Lamb, *Annual Report on the Forest Administration of Sierra Leone 1958/1959* (Government Printer, 1960).
15. A.F.A. Lamb, *1956 Report on the Forest Administration* (Government Printer, 1957); Lamb, *Annual Report on the Forest Administration of Sierra Leone 1958/1959*.
16. A.F.A. Lamb, *Report on the Forest Administration of Sierra Leone 1957* (Government Printer, 1959).
17. J. E. F. Douay, 'Yemane-Gmelina arborea (Roxb.) A Monograph, Including a Preliminary Test for Suitability as Matchwood Carried Out on Plantation-Grown Material from Sierra Leone', MSc (Imperial Forestry Institute, Oxford University, Oxford, 1954); A.K. Nicol, *Report on the Forest Administration of Sierra Leone for the Year 1954* (Government Printer, 1955); Sawyerr, 'National Report on Man-made Forests (Sierra Leone)'.
18. D.H. Hodgson, *Annual Report of the Forest Department for the Year 1950* (Government Printer, 1952); R.S. Pelly, *Annual Report on the Forest Administration for the Year 1951* (Government Printer, 1952).
19. *The Daily Mail*, 'From Man-Made Forest', 12 November, 1959, 1.

20. R. Plumptre, 'Obituary – Dr. H.C. Dawkins MBE', *Forestry* 65(4) (1992), 479–80.
21. H.C. Dawkins, *Report to the Forestry Division of the Ministry of Natural Resources: On Problems of Natural Regeneration, Plantation and Research* (Commonwealth Forestry Institute, 1965), 1.
22. Dawkins, *Report to the Forestry Division of the Ministry of Natural Resources*, 26.
23. C. Swabey, 'Visit of Forestry Adviser to Sierra Leone – August 1958', Tours of Forestry Adviser 1957–1962: RHA – MSS Brit Emp s. 459.
24. Hodgson, *Annual Report on the Forest Administration for the Year 1949.*
25. J.A. Helms, *Society of American Foresters* (Society of American Foresters, 1988).
26. See R. Drayton, *Nature's Government: Science, Imperial Britain, and the 'Improvement' of the World* (Yale University Press, 2000); H. Tilley, *Africa as a Living Laboratory: Empire, Development, and the Problem of Scientific Knowledge, 1870–1950* (University of Chicago Press, 2011).
27. C.E. Lane-Poole, *Report on Forests of Sierra Leone* (London: Waterlow & Sons, 1911); Sawyerr, 'National Report on Man-made Forests (Sierra Leone)'; *The Commonwealth Forestry Review*, 'Sierra Leone' 50 (1971), 144–45.
28. J. Fairhead and M. Leach, *Reframing Deforestation: Global Analyses and Local Realities with Studies in West Africa* (Routledge, 1998).
29. Colonial Secretary. Communication of Forest Reserves and re-afforestation – 26 May 1918: SLNA – AG/45/17.
30. *The Empire Forestry Journal*, 'REVIEW: Sierra Leone: Annual Report of the Forest Department for the Year 1930' 11 (1930), 150–51.
31. E. MacDonald, Government Communicaiton on Re-organisation the Forestry Department: 10 April, 1929: SLNA A/43/29; E. MacDonald, *Sierra Leone Annual Report of the Forest Department for the Year 1929* (Government Printer, 1930).
32. Bryant, 'The Rise and Fall of Taungya Forestry', 21–26; Menzies, 'Three Hundred Years of Taungya', 361–76.
33. P.K.R. Nair, *An Introduction to Agroforestry* (Kluwer Academic Publishers, 1993).
34. Sawyerr, 'National Report on Man-made Forests (Sierra Leone)', 16.
35. Nair, *An Introduction to Agroforestry.*
36. E. MacDonald, Taungya Plantations. Advances to Farmers Communication to Colonial Treasurer, Freetown – 1 April, 1932: SLNA A/29/32.
37. E. MacDonald, *Sierra Leone Annual Report of the Forest Department for the Year 1931* (Government Printer, 1932); M.A. Young, Communication to Lord Passfield (Government House) – 27 July, 1929: SLNA – A/43/29.
38. MacDonald, Communication to C.V. Wallace: Forestry Department handing over notes, 1929: SLNA – CSO 2/8; E.P. Stebbing, *The Forests of West Africa and the Sahara: A Study of Modern Conditions* (W & R Chambers, 1937).

39. Forest Department Handing Over Notes: November, 1935. Conservator of Forest to Assistant Conservator (Mr C.V. Wallace), SLNA A/192/29.
40. D.H. Hodgson, *Report on the Forest Administration of Sierra Leone for the Year 1948* (Government Printer, 1949); E. Vaclav, 'Trees of Forest Plantations in Sierra Leone', *Silvaecultura Tropica et Subtropica* 6 (1978), 65–69.
41. MacDonald, Communication to C.V. Wallace: Forestry Department handing over notes, 1929: SLNA – CSO 2/8; Stebbing, *The Forests of West Africa and the Sahara*.
42. Stebbing, *The Forests of West Africa and the Sahara*; also see E.P. Stebbing, 'The Man-Made Desert in Africa: Erosion and Drought', *Journal of the Royal African Society* 37(46) (1938), 3–40.
43. Stebbing, *The Forests of West Africa and the Sahara*, 163.
44. W.M. Robertson, 'Sierra Leone', *News Bulletin of Empire Forest Departments for the First Half of 1937* (1937), 25–29.
45. E. MacDonald, Notes on the Committee on land deterioration in Sierra Leone, 1938: SLNA – A/42/38.
46. W.D. MacGregor, *Report on the Forest Administration of Sierra Leone for the Year 1944* (Government Printer, 1945); Pelly, *Annual Report on the Forest Administration for the Year 1951*.
47. H. Maughan-Brown, 'Sei Bush Belt in Sierra Leone', *Farm and Forest* 4(1) (1943), 8.
48. Maughan-Brown, 'Sei Bush Belt in Sierra Leone', 8–9; Lamb, *1956 Report on the Forest Administration*; R.S. Pelly, *Annual Report on the Forest Administration for the Year 1953* (Government Printer, 1955).
49. Hodgson, *Report on the Forest Administration of Sierra Leone for the Year 1948*.
50. John Wools, Personal Communication, 2011.
51. W.D. MacGregor, *Sierra Leone Forestry Report for the Year 1941* (Government Printer, 1942); W.D. MacGregor, *Report on the Forest Administration of Sierra Leone for the Year 1946* (Government Printer, 1947); Maughan-Brown, 'Sei Bush Belt in Sierra Leone', 8–9; Hodgson, *Report on the Forest Administration of Sierra Leone for the Year 1947*; Hodgson, *Annual Report on the Forest Administration for the Year 1949*; Pelly, *Annual Report on the Forest Administration for the Year 1953*.
52. R.A. Cline-Cole, 'Towards an Understanding of Man-Firewood Relations in Freetown (Sierra Leone)', *Geoforum* 15(4) (1984), 583–94; P.G. Munro and G.A. van der Horst, *The Domestic Trade in Timber and Fuelwood Products in Sierra Leone: Current Dynamics and Issues* (FAO/EU, 2012).
53. Hodgson, *Report on the Forest Administration of Sierra Leone for the Year 1948*, 98.
54. K.S. Amanor, 'Farmers, Forestry and Fractured Environmentalism in Ghana's Forest Zones' in R.A. Cline-Cole and C. Madge (eds), *Contesting Forestry in West Africa* (Ashgate, 2001), 307–21; S. Jha, 'Radical Politics and Environmentalism against Taungya in Dooars', *Economic & Political*

Weekly 47(1) (2012),112–18; R.P. Neumann, *Imposing Wilderness: Struggles over Livelihood and Nature Preservation in Africa* (University of California, 1998); P. Vandergeest and N.L. Peluso, 'Empires of Forestry: Professional Forestry and State Power in Southeast Asia, Part 2', *Environment and History* 12(3) (2006), 359–93; P. von Hellermann, 'Things Fall Apart? Management, Environment and Taungya Farming in Edo State, Southern Nigeria', *Africa* 77(3) (2007), 371–92; N.L. Peluso, 'The History of State Forest Management in Colonial Java' *Forest & Conservation History* 35(2) (1991), 65–75.

55. Bryant, 'The Rise and Fall of Taungya Forestry', 21.
56. Bryant, 'The Rise and Fall of Taungya Forestry', 21.
57. Vandergeest and Peluso, 'Empires of Forestry', 359–93.
58. Bulmer, *Report on the Forest Administration of Sierra Leone 1959/1960*; I. Paul, *Report on the Forest Administration of Sierra Leone 1960/61* (Freetown: Government Printing Department, 1961); Lamb, *Annual Report on the Forest Administration of Sierra Leone 1958/1959*; J.E.D. Fox, 'Natural Regeneration of the Kambui Hills Forest in Eastern Sierra Leone. Part I. Ecological Status of the Lophira/Heritiera Rain Forest', *Tropical Ecology* 11 (1970), 169–85; J.E.D. Fox, 'Natural Regeneration of the Kambui Hills Forest in Eastern Sierra Leone. Part II. Methods of Silvicultural Treatment and Analysis', *Tropical Ecology* 12 (1971), 1–23; J.E.D. Fox, 'Natural Regeneration of the Kambui Hills Forest in Eastern Sierra Leone. Part III. The Results of Silvicultural Treatment and Management Consideration', *Tropical Ecology* 13 (1972), 139; Peter Savill, Personal Communication, 2011.
59. I. Paul, 'Progress Report 1955–60 by the Forestry Department of Sierra Leone', British Commonwealth Forestry Conference, 1962.
60. W.D. MacGregor, *Sierra Leone Forestry Report for the Year 1942* (Government Printer, 1943).
61. Lamb, *1956 Report on the Forest Administration*; Lamb, *Report on the Forest Administration of Sierra Leone 1957*; Paul, *Report on the Forest Administration of Sierra Leone 1960/61*.
62. Paul, *Report on the Forest Administration of Sierra Leone 1960/61*, 61.
63. Lamb, *Annual Report on the Forest Administration of Sierra Leone 1958/1959*; Paul, *Report on the Forest Administration of Sierra Leone 1960/61*; J.S. Sawyerr, 'The Use of Melaleuca Leucodendron in Mangrove Soil Reclamation in Sierra Leone', paper presented at the Fifth Meeting of CROACUS, Freetown, 1963.
64. Hodgson, *Report on the Forest Administration of Sierra Leone for the Year 1948*.
65. J.S. Sawyerr, *Report on the Forest Administration of Sierra Leone for the Year April, 1963 to March, 1964* (Government Printer, 1965).
66. Paul, *Report on the Forest Administration of Sierra Leone 1960/61*; Sawyerr, *Report on the Forest Administration of Sierra Leone for the Year April, 1963 to March, 1964*; Sawyerr, 'National Report on Man-made Forests (Sierra Leone)'.

67. John Woolls, Personal Communication, 2011.
68. W.M. Ciesla and E. Donaubauer, *Decline and Dieback of Trees and Forests: A Global Overview*, vol. 120 (Food & Agriculture Organization of the UN (FAO), 1994).
69. Lamb, *1956 Report on the Forest Administration.*
70. Hodgson, *Report on the Forest Administration of Sierra Leone for the Year 1947*; Hodgson, *Report on the Forest Administration of Sierra Leone for the Year 1948.*
71. Fox, 'The Growth of *Gmelina Arborea*', 138–44.
72. Tinsley and Barrett, *Report on a Visit to Sierra Leone to Investigate the 'Die-Back' Condition of Gmelina Arborea* (University of Oxford, Commonwealth Forestry Institute, 1965).
73. Lamb, *1956 Report on the Forest Administration*; Lamb, *Report on the Forest Administration of Sierra Leone 1957.*
74. Tinsley and Barrett, *Report on a Visit to Sierra Leone.*
75. M.B. Feika, *Report on the Forest Administration of Sierra Leone 1963/1964 to 1973/74.*
76. Dawkins, *Report to the Forestry Division of the Ministry of Natural Resources*; Lamb, *1956 Report on the Forest Administration*; Tinsley and Barrett, *Report on a Visit to Sierra Leone.*
77. Tinsley and Barrett, *Report on a Visit to Sierra Leone.*
78. Fox, 'The Growth of *Gmelina Arborea*', 138–44; J.S. Sawyerr, *Ten-Year Plan of Forestry Development for Sierra Leone: 1964/5 -1973/1974* (Ministry of Natural Resources, 1964).
79. Bulmer, *Report on the Forest Administration of Sierra Leone 1959/1960.*
80. Sawyerr, *Ten-Year Plan of Forestry Development for Sierra Leone*; Paul, 'Progress Report 1955–60'.
81. Sawyerr, 'The Use of Melaleuca Leucodendron'; Sawyerr, *Report on the Forest Administration of Sierra Leone for the Year April, 1963 to March, 1964*; Sawyerr, 'National Report on Man-made Forests (Sierra Leone)'.
82. Feika, *Report on the Forest Administration of Sierra Leone 1963/1964 to 1973/74.*
83. P. Cox, G. Hemmant, F.M. Oliphant, R.S. Troup, A. Hill and E.P. Stebbing, 'The Encroaching Sahara: The Threat to the West African Colonies', *Geographical Journal* 85(6) (1935), 510.
84. Stebbing, *The Forests of West Africa and the Sahara*, 3.
85. Cox et al., 'The Encroaching Sahara', 519–24.
86. Cox et al., 'The Encroaching Sahara', 519–24.
87. Stebbing, *The Forests of West Africa and the Sahara.*
88. Stebbing, *The Forests of West Africa and the Sahara*; Stebbing, 'The Man-Made Desert in Africa', 3–40.
89. Sierra Leone Forest Department (1935). Exotics – Statement Submitted by the Forest Department, Sierra Leone, West Africa.

90. See G. Hiemstra-van der Horst and A.J. Hovorka, 'Fuelwood: The 'Other' Renewable Energy Source for Africa?', *Biomass and Bioenergy* 33(11) (2009), 1605–16.
91. P.G. Munro, G. van der Horst and S. Healy, 'Energy Justice for All? Rethinking Sustainable Development Goal 7 through Struggles over Traditional Energy Practices in Sierra Leone', *Energy Policy* 105 (2017), 635–41.
92. Munro, Van der Horst and Healy, 'Energy Justice for All?', 635–41.
93. J.C. Ribot, 'A History of Fear: Imagining Deforestation in the West African Dryland Forests', *Global Ecology and Biogeography* 8(3–4) (1999), 291–300.
94. ECOWAS, *Proceedings of the Economic Community of West African States (ECOWAS) Energy Symposium*. Energy For Survival (ECOWAS, Trans.) (Freetown, 1982), 25.
95. O.R. Davidson, *Energy Use Patterns, Sierra Leone* (International Development Research Centre, 1985), 1.
96. R.G. Johnson, M. Kandeh, A. Jalloh, G.C. Nelson, and T.S. Thomas, 'Sierra Leone', in A. Jalloh, G.C. Nelson, T.S. Thomas, R. Zougmoré and H. Roy-Macualey (eds), *West African Agriculture and Climate Change: A Comprehensive Analysis* (IFPRI, 2013), 329.
97. For example, see J.E. Arnold, G. Köhlin and R. Persson, 'Woodfuels, Livelihoods, and Policy Interventions: Changing Perspectives', *World Development* 34(3) (2006), 596–611; R.A. Cline-Cole, H.A.C. Main, J.E. Nichol, 'On Fuelwood Consumption, Population Dynamics and Deforestation in Africa', *World Development* 18(4) (1990), 513–27; G. Leach and R. Mearns, *Beyond the Woodfuel Crisis: People, Land and Trees in Africa* (Earthscan, 1988); G.A. Hiemstra-Van der Horst and A.J. Hovorka, 'Reassessing the 'Energy Ladder': Household Energy Use in Maun, Botswana', *Energy Policy* 36(9) (2008), 3333–44; Hiemstra-van der Horst and Hovorka, 'Fuelwood',1605–16.
98. Hiemstra-van der Horst and Hovorka, 'Fuelwood', 1605–16.
99. For example see E.K. Alieu, *Country Perspective – Sierra Leone* (Commonwealth Forestry Association, 2011); Davidson, *Energy Use Patterns*; FAO, *Sierra Leone: BEFS Country Brief* (Food and Agriculture Organization of the United Nations (FAO), 2013); GoSL, *Sierra Leone Forestry Policy 2010* (Government of Sierra Leone, 2010); J. Kamara, *Firewood Energy in Sierra Leone – Production, Marketing, and Household Use Patterns* (Verlag Weltarchiv, 1986).
100. For an overview of CARE tree planting in West Africa, see J. Sumberg and M. Burke, 'People, Trees and Projects: A Review of CARE's Activities in West Africa', *Agroforestry Systems* 15(1) (1991), 65–78.
101. See E.K. Alieu, 'Agro-forestry Practices in Sierra Leone', unpublished paper (1983); E.K. Alieu, 'The Birth of the First Village Forestry Association in Sierra Leone', *Y Coedwigwr* 40 (1989), 2-4; E.K. Alieu, *Forestry Outlook Studies in Africa (FOSA) – Sierra Leone* (Ministry of Natural Resources and

Tourism, 2001); P.D. Cummings, 'Community Forestry – Prospects For Sierra Leone', MSc Environmental Forestry (University of Bangor, Bangor, 1990); S.E. During, *Fish Smoking Trials with Acacia Tree Species as Fuel Wood* (Freetown: GTZ, 1990); T.E.A. Lansana, *A Study of the Firewood Species Used for Fish Processing in the Tombo Village* (GTZ, 1986); R.V. Potter, L. Danso and P.D. Palmer, *Community Participatory Forestry for Fuelwood Production in Western Area, Sierra Leone: Report of an Evaluation Mission* (FAO, 1990); P. Tomasi, *Starting Community Forestry in the Western Region of Sierra Leone through Extension and Cooperation with Non-governmental Organizations* (FAO, 1987).

102. E.K. Alieu, 'Building on Local Foundations: Enhancing Local Community Support for Conservation', *Unasylva* 61 (2010), 22–27.
103. P. Munro, 'Deforestation: Constructing Problems and Solutions on Sierra Leone's Freetown Peninsula', *The Journal of Political Ecology* 16 (2009), 104–24.
104. C. Oxby, 'Changes in Shifting Cultivation in Africa', *Unasylva* 37 (1985).
105. P.C. Goswami and M. Hoskins, *Assistant to Local Community Forestry* (Rome: FAO, 1980); M. Illes, P.S. Savill and G. Koker, *Gola Forest Reserves Sierra Leone, Interim Management Plan* (Sandy: RSPB, 1993).
106. Alieu, 'Agro-forestry Practices in Sierra Leone'; A.P. Koroma, 'Taungya in Sierra Leone', in L.H. MacDonald (ed.), *Agro-forestry in the African Humid Tropics* (The United Nations University, 1981), 67–69; Illes, Savill and Koker, *Gola Forest Reserves Sierra Leone.*
107. M. Leach and R. Mearns, 'Environmental Change and Policy', in M. Leach and R. Mearns (eds), *The Lie of the Land: Challenging Received Wisdom on the African Environment* (James Currey, 1996), 1–33.
108. Sawyerr, *Ten-Year Plan of Forestry Development for Sierra Leone*, 6; R.P. Neumann, *Making Political Ecology* (Hodder Arnold, 2005).
109. Alieu, *Forestry Outlook Studies in Africa.*
110. S. Matthews and K. Brand, *Africa Invaded: The Growing Danger of Invasive Alien Species* (GISP, 2004).
111. B.N. Kamara, *Biodiversity: Strategic Action Plan* (Freetown Government of Sierra Leone, 2003), 56.
112. B.N. Kamara, *CBD Sierra Leone – Third National Report* (Freetown Government of Sierra Leone, 2008).
113. Dawkins, *Report to the Forestry Division of the Ministry of Natural Resources*; A.K. Nicol, *Annual Report on the Forest Administration of Sierra Leone for the Year 1952* (Government Printer, 1953); A. Sanon, P. Martin, J. Thioulouse, C. Plenchette, R. Spichiger, M. Lepage and R. Duponnois, 'Displacement of an Herbaceous Plant Species Community by Mycorrhizal and Non-mycorrhizal Gmelina Arborea, an Exotic Tree, Grown in a Microcosm Experiment', *Mycorrhiza* 16 (2006), 125–32.
114. R.A. Cline-Cole, 'The Urban Fuel Plantation in Tropical Africa: A Case for Re-evaluation', *Land Use Policy* 7(4) (1990), 323–36.

115. K.S. Amanor, 'The Symbolism of Tree Planting and Hegemonis Environmentalism in Ghana', paper presented at Changing Perspectives on Forests: Ecology, People and Science/Policy Processes in West Africa and the Carribean, University of Sussex, 2001.
116. N. Mueller-Hirth, 'If You Don't Count, You Don't Count: Monitoring and Evaluation in South African NGOs', *Development and Change* 43(3) (2012), 649–70.
117. D.G. Thomas, Forest Department reafforestation scheme for Bo, 1937: SLNA - A/51/37.
118. Sawyerr, *Ten-Year Plan of Forestry Development for Sierra Leone*, 6.
119. Neumann, *Making Political Ecology.*
120. Vandergeest and Peluso, 'Empires of Forestry', 359–93.
121. Munro and Van der Horst, *The Domestic Trade in Timber and Fuelwood Products.*

CHAPTER 4

Exploitation

In December 1939, the Scotsman William Douglas MacGregor was appointed as the Head of the Forestry Department in Sierra Leone. He came to the post after fifteen years of working for the colonial forestry service in Nigeria.[1] It was an inauspicious time to take the reins of forestry in the colony. World War II had just broken out. As MacGregor observed, shortly after his arrival in Sierra Leone, 'War clouds overshadow everything and the Forestry Development has virtually come to a standstill.'[2] The last time a world war had broken out, some two and half decades earlier, forestry activities in Sierra Leone had come to halt, as all forestry staff, including the then Head of the Department, Charles Lane-Poole, had been seconded to wartime activities, a situation that set the basis for a poorly funded and haphazard approach to colonial forestry in Sierra Leone during the first half of the twentieth century.

Initially, it seemed that forestry might suffer a similar fate during World War II; after its outbreak, two forestry officers were released for military and naval duties, a third was diverted to wartime censorship duties, and for a period in 1940, MacGregor was the only senior forestry officer left at the Department.[3] During that same year, almost all on-the-ground forestry activities were terminated, and the Department was essentially relegated to taking on an advisory role.[4] Given its history of struggles and lacklustre achievements, it would not have been all that surprising if the War resulted in the death knell of colonial forestry in Sierra Leone. It may well have, if it were not for William MacGregor's tenacity. MacGregor, early in his tenure, declared that 'so far we have nibbled at timber exploitation', boasting that he would develop a new exploitation programme that would 'be revolutionary'.[5]

It was an ambitious statement, but subsequent actions would match this rhetoric, and arguably no single individual has done more than MacGregor to change the shape of the commercial forestry sector in Sierra Leone.

In June 1941, MacGregor sent two letters to the Colonial Secretary in Freetown that both demanded the same thing – the establishment of a Forestry Corps in Sierra Leone. The idea was apparently devised over breakfast with his forest exploitation subordinate Thomas E. Edwardson. A forestry labour corps would be established in Sierra Leone along army lines that would oversee timber exploitation during the War.[6] All MacGregor needed was to secure funding for the initiative, and the War offered opportunities for this. His first letter laid out the 'realities' of commercial forestry in Sierra Leone: limited staff, the loss of sawyers to army recruitment and an overall shortage of tools.

Therefore, the formation of the Forestry Corps, under the newly created Defence (Compulsory) Service Regulations, was the only way to guarantee a constant timber supply in Sierra Leone – an imperative during wartime conditions.[7] As MacGregor pleaded:

> The point about raising a Forestry Corps is that we would have a disciplined body of forest workers who would be organised on the basis of maximum production. Timber production would not be at the mercy of recruitment campaigns for other purposes or the casual efforts of private enterprise.[8]

His second letter, sent a month later, had a more desperate tone.

> Local private enterprise is incapable of organising and expanding the timber industry on a self-sufficiency basis. Most of the contractors (i.e. employers of sawyers) are either inefficient, apathetic, lacking in capital, or, because of their efforts to screw the maximum profit out of industry, they are unable to main a permanent organisation. Sawyers work spasmodically, and their outturn is low in quality and quantity.[9]

The latter claim was likely hyperbole, as just two years earlier a colonial forestry expert in the area of timber exploitation had travelled to Sierra Leone to observe the local pit-sawyers, noting that 'The quality and accuracy of the sawing is first class and the equal of the best hand-sawing I have seen in any tropical country. The sawyers have a long swinging stroke and it is difficult to believe they are not to the manner born.'[10] This second letter, nevertheless, was effective. MacGregor got his Forestry Corps. There was, however, one important distinction, in that MacGregor had wanted the Forestry Corps to come under the jurisdiction of the Forestry Department, but instead the Colonial Secretary decided that they were to come under the jurisdiction of the War Office, and therefore their operations were to be external, yet complementary, to the Forestry Department's own operations. The Royal Engineers Forestry Corps arrived in Sierra Leone in 1943, having previously been based in Sherwood Forest in England, and remained in the colony until the end of the War.[11]

In addition to securing the Forestry Corps, MacGregor also managed to source a number of sawmills. In a seemingly competitive West African forestry spirit, he bragged about this achievement in *The Empire Forestry Review* journal:

> Nigeria has beaten us by a short head, but I think Sierra Leone probably has the second and third [saw]mill of the type under discussion. A forest mill (read Pilot mill) was proposed for one of our up-country forests in 1941 … There were inevitable delays, but in 1942 the timber situation was such that our ideas had to be enlarged. Machinery arrived early 1943, and as I write, one mill is just on the point of going into production and the second mill will be in operation by July. A third is mooted, but it will be a much smaller affair and really portable.[12]

Four sawmills were secured in total. The two biggest were to be owned and operated by the Army Corps – one at River No. 2 (a few kilometres south of Freetown), and the second at Kasewe Forest Reserve, along the railway line, around 100 kilometres east of Freetown. The Forestry Department also managed to secure a smaller sawmill for its operations in the eastern city of Kenema, while another sawmill was installed in Freetown, with the specific purpose of sawing mangrove wood to produce fuel for locomotives.[13] As MacGregor declared, 'The year 1943 has seen the turning point in the history of the Department, stabilising, it is hoped, an active forest policy. African and European personnel have been increased, sawmills have been built and the utility value of forest has been keenly appreciated.'[14] MacGregor continued to promote timber exploitation, and even before the War had ended, he negotiated an agreement to ensure that ownership of the Forestry Corps sawmills would pass over to the Forestry Department after the War finished. MacGregor had secured his forest exploitation 'revolution'.

Soon after the War, in 1947, MacGregor retired from the colonial forestry service, taking up a lecturing position at the University of Aberdeen.[15] This was not before his great achievement during the War period was recognised by higher colonial authorities with the award of a Commander of the Most Excellent Order of the British Empire (CBE) in 1945.[16] MacGregor had overseen a transformation of forestry in Sierra Leone, from a peripheral resource department that was a drain on the Colonial Treasury, to a resource extraction industry that was producing enough timber for domestic needs (there were even some timber exports to The Gambia in 1947) and a forest utilisation (exploitation) branch that was turning over a small profit each year. Between 1942 and 1944 alone, the amount of timber being produced in Sierra Leone increased twenty-fold.[17] World War II had been used as a strategic event for securing future commercial forestry in Sierra Leone.[18] However, it did not end there, as 'sawmilling, which was an outcome of wartime necessity ha[d] come to stay'.[19]

The William MacGregor story brings into focus some of the vagaries of forest exploitation in Sierra Leone during the colonial era. Evidently, the emergence of Sierra Leonean forestry exploitation activities was not a monolithic programme devised by colonial government, but rather it arose out of the manoeuvring by the Head of the Forestry Department, who was able to take advantage of the political economy of World War II. A poorly funded and lacklustre Forestry Department was potentially only able to survive thanks to the changing international geopolitics and strategic local lobbying. Overall, as the rest of this chapter explores, this was ultimately a central narrative for forestry exploitation in Sierra Leone; its realisation continued to be a precarious endeavour, and its success and operations were heavily mediated by changing political economies and protagonist individuals.

Exploitation Tensions

Forest exploitation, in complement to the reservation and plantation programmes, completed the logic of colonial forestry. Reservation secured land and forests; plantations ensured that many reserves were filled with 'useful' economic trees; and exploitation subsequently focused on transforming trees into commodities (i.e. timber) and producing profit for the colonial government. There was, however, ongoing tension surrounding *when* Sierra Leone was going to be ready for large-scale exploitation to commence and ultimately how central timber exploitation should be, and this was played out between Sierra Leone's two earliest colonial foresters – Arthur Unwin, who was sent to the colony in 1908 to evaluate its forestry potential, and Charles Lane-Poole, who was sent in 1910 to set up the colony's Forestry Department.

Unwin saw immediate profit in Sierra Leone's forests and recommended the installation of sawmills along with an immediate start to large-scale exploitation of Sierra Leone's forest. Such a position was perhaps understandable given his background.[20] Unwin had worked in Nigeria as a colonial forester for many years[21] and was likely appreciative of the profit-orientated goals of the colonial project – for forestry to be viable, it needed to provide timber and revenue. Lane-Poole, in contrast, was grounded in the 'science' of forestry (he was a formally trained forester, Unwin was not) and had very little previous on-the-ground colonial work experience.[22] Thus, while Unwin imagined that 'the Department will be very largely remunerative even from the beginning',[23] Lane-Poole argued that 'all that can be said is that there will be little or no revenue for some five to six years. The work of the Department will up to that time be confined to conserving [i.e. forest reservation work]'.[24] Even their reading of the landscape differed: in relation to a small island near Freetown, Unwin noted, 'the wooded hills of the Banana Islands should not be forgotten, as they are covered with a valuable forest',[25] while Lane-Poole countered, 'The Banana Islands should be excluded from [becoming a] reserve as there is no timber of any value there.'[26] For Unwin, forestry was beholden to the colonial government; for Lane-Poole, it was beholden to forestry scientific ideals. William MacGregor later on summed up these conflicting approaches neatly:

> There are two schools of thought about forest finance. One holds that a forest department should be considered as a long-term investment for the stabilisation of rural prosperity, and that the absence of immediate direct revenue should not affect continuity of investment. The other school would put forestry on an immediate revenue earning basis.[27]

Thanks to Lane-Poole's initial tutelage, the early years of the colonial Forestry Department were focused on his more 'conservationist' approach to forestry. It was not an approach that was anti-exploitation by any means but rather one that saw broader colonial aims (i.e. creating profitable colonies) as being subordinate to 'scientific' forestry management. Sierra Leone needed a forest estate before it could enact an exploitation programme, hence the initial narrow focus on forest reservation creation.[28] Indeed, according to Forestry Department annual reports, all formal timber exploitation activities during the colonial era in Sierra Leone occurred in Forest Reserves and Protected Forests. The forest estate provided the basis for timber exploitation.

Given this incremental approach to exploitation, the Sierra Leonean Government and Freetown's urban populace were heavily dependent on imports to fulfil their timber needs during the first three decades of the twentieth century. Timber during this period was mainly imported from the United States and the Baltic Sea region.[29] Opportunities for Departmental timber exploitation during the period were largely limited to occasionally supplying sleepers to the Railway Department and the hope that future mining developments, near Forest Reserves, might create some strategic demand for timber.[30] The Forestry Department's operations were, therefore, almost completely reliant on funding from the Colonial Government. For example, in 1924, the revenue received from timber licences was reported as being £50, while the entire Department's expenditure was £6,941.[31] It is unsurprising that the long-term economic viability of its operations was soon to come under the scrutiny of the Colonial Government.

In 1930, in response to a Colonial Government ultimatum, the Forestry Department was manoeuvred towards (more profitable) exploitation activities, and Lane-Poole's legacy – an overt policy focus on reservationism – began to fade. MacGregor usefully narrated this change:

> Up to 1930 the Department was concerned with saving, by reservation, the remnants of the country's forest resources – no revenue was earned or expected, but when the 1931 depression came expenditure was drastically reduced and Government instructed the Conservator to stop reservation and start exploitation.[32]

The then head of the Forestry Department, Eric MacDonald, personally undertook enumeration surveys in 1930 to identify forest areas that were suitable for exploitation.[33] These surveys took place in the Kambui Hills, one of Sierra Leone's earliest established Forest Reserves (see

Chapter 2) – an area that had frequently been identified as a potential site for exploitation due to its forest stocks and its proximity to Kenema (a major urban centre) and the railway line.[34] In 1933, MacDonald devised an experiment in the Kambui Hills Forest Reserve – the Department organised for forty trees to be cut and logs to be dragged to the edge of the Reserve (about one kilometre from the railway), where they expected to sell them to local contractors.[35] The experiment failed – not a single log was sold despite them being below market price – and MacDonald was incensed:

> It is possibly a little difficult for anyone who has not experienced it to realise how backward the timber industry is in Sierra Leone. For three weeks now, some 40 good logs of desirable species have been lying in a depot only ¾ mile from Kenema Rail Station, and we have not yet succeeded in getting an offer of any kind for them, although there is a notorious shortage of native timber. In such circumstances there is a considerable risk of local criticism to the effect that enumeration surveys are 'useless'.[36]

MacDonald, nevertheless, persevered. The following year, 1935, he set up gangs of local pit-sawyers, who managed to produce a modest amount of timber for the Sierra Leonean domestic market – accounting for around 15 per cent of the year's domestic timber needs.[37] Some minor adjustments were also made to the Forestry Ordinance, which included an increase in forestry fees and royalties as well as the reopening of the Forestry Office in Kenema, which was closed in 1932.[38] MacDonald also organised for the research, and subsequent purchase, of a Witte Saw from Liverpool[39] – however, this proved to be 'quite useless for the purposes for which it was bought', as its engine was inadequate for sawing the large diameters of tropical timbers.[40] During the late 1930s, the programme slowly expanded. Shimbeck huts – free accommodation for pit-sawyers – were established in Kenema, and a supply deal was negotiated with the Railway Department. MacDonald also managed to get a utilisation officer position underwritten through the Colonial Development fund.[41] Even an entire rail locomotive was built from local timbers, which was to be used on the railway lines that stretch from Freetown to Kenema and beyond.[42] Progress was steady; however, the timber produced by the Department was still only enough to supply a very small percentage of the domestic market – with the bulk of commercial timber still being imported into Sierra

Leone. Pit-sawing was too limited in terms of its potential production volume.[43]

Given these modest outputs, obtaining a large sawmill soon became a major objective for the Department. The Colonial Government was cautious about making such an investment, and therefore invited Major Wilkins F. Chipp, of the Malaysia Forestry Service (an expert in saw-milling), to evaluate the country's commercial forestry potential. Chipp spent just thirteen days in Sierra Leone, inspecting the active exploitation programme and a handful of Forest Reserves (i.e. Kambui Hills, Gola). He was particularly impressed with the local pit-sawyers.[44] The rest of his evaluation, however, was largely negative. The Major concluded that mechanical logging of the Kambui Hills was not 'an economic proposition' and noted that transportation was a major issue – most areas were inaccessible and rail freight was expensive. He deemed that the largely inaccessible Gola North Forest Reserve was the only area with the potential for being exploited by mechanical methods.[45] *The Empire Forestry Journal*, drawing upon Chipp's findings, provided a frank assessment of how the Forestry Department should proceed with its exploitation programme:

> It would appear that there has been considerable improvement in the standard of hand-sawing in recent years, due to experimental exploitation at Kenema. It is probable that the next step should be encouragement of the local timber trade and not this anxiety to introduce mechanical methods of exploitation that is so stressed in the [annual Sierra Leone forestry] report.[46]

The sudden outbreak of World War II, and the arrival of the opportunistic William MacGregor, however, meant that this advice was roundly ignored.[47]

Exploitation Revolution

The impact of MacGregor's initiatives during World War II – securing a forestry corps and a number of sawmills – fundamentally changed the commercial forestry sector. At the end of the War, as had been planned, the two timber sawmills that had been operated by the Army were transferred to the Forestry Department. Operations at the River No. 2 sawmill were immediately ceased, and the sawmill was then dismantled,

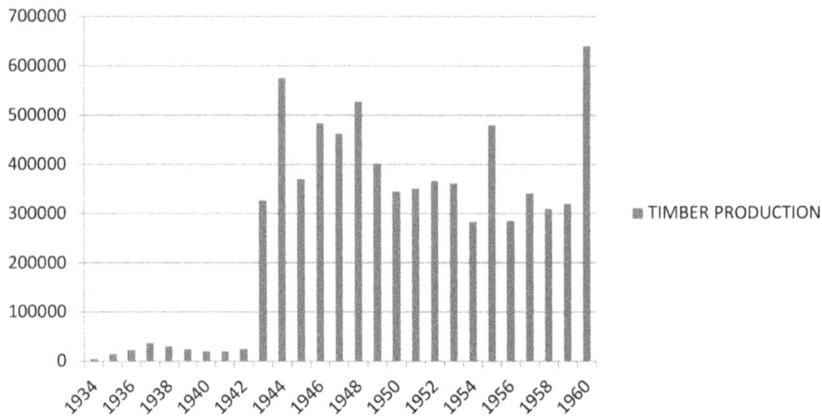

Figure 4.1 Commercial timber production in Sierra Leone by volume – 1934 to 1960. Data sourced from Forestry Department Annual Reports. Created by the author.

transferred and re-assembled in Kenema. Sawmill operations continued at Kasewe Forest Reserve until 1949, and then it too was dismantled and transferred to Kenema.[48] As such, Kenema was transformed into the 'timber capital of the nation',[49] and over the next couple of decades the urban growth of the city was heavily driven by the expanding forestry industry.[50] Former Forestry Corps army staff also chose to stay in Sierra Leone and work for the Forestry Department's newly created Utilisation Branch (i.e. a branch focused on timber exploitation).[51] The Forestry Department now had multiple sawmills, a cohort of experienced exploitation staff and was even sending experts to the British Colony of The Gambia to help develop its commercial forestry sector.[52] Initial exploitation activities after the War were focused on the nearby Kambui Hills Forest Reserve, while the construction of an important bridge and road to the Gola North Forest Reserve by the Colonial Government was keenly anticipated by the Department until its completion in 1960.[53] Thanks to these initiatives, the high timber production yields achieved during the War were maintained until Sierra Leone's independence in 1961 (see Figure 4.1). Domestic production was largely able to meet domestic needs, with any imports coming into the country during this period being largely to fulfil specialist timber needs (e.g. for furniture making) (see Figure 4.2).[54]

The maintenance of production levels was nevertheless always a challenge. Staff shortages continued to be a problem, electricity supply for the

Figure 4.2 Percentage share of domestic timber production in Sierra Leone's timber market. Data sourced from Forestry Reports, *The Government Gazette* and the *Sierra Leone Trade Journal*. Created by the author.

mill was erratic, rail freight charges were high and rail wagon space was limited.[55] This was in part due to the design of the train line, which was on a very small gauge that was poorly designed for the context in which it was being used.[56] The 'diamond mining rush' that occurred in Sierra Leone in the 1950s also produced problems and opportunities. Many forestry labourers left the Department to partake in 'illicit' diamond mining,[57] yet at the same time the influx of money from the diamond business gave rise to a greater market for timber demand.[58] All in all, the exploitation branch of the Forestry Department endured and slowly increased its sales, and although never quite reaching its annual production targets, it did manage to diversify its marketing portfolio by developing a workshop where high-quality carpentry items were produced. This included producing roofing shingles for houses and then, later, fully fabricated houses (see Figure 4.3).[59] A wide array of furniture was also produced whose quality received much praise. The Department was supplying 'schools, colleges, hospitals, government offices and private individuals with furniture' and during this period was employing around 600 people.[60] It received an important windfall in 1960, when it secured the contract (against much overseas competition) to build and supply the furniture for the newly constructed Paramount Hotel in Freetown – the most prominent hotel in Sierra Leone at the time.[61] This success was

Figure 4.3 Prefabricated houses (R.S. Pelly, 'Forest Industries – Sierra Leone', *The Empire Forestry Review* 29 (1950), 351–60).

reported in the *Daily Mail*, the country's widest circulating newspaper, with a special feature article containing photographs and complements focused on the Forestry Department's high quality furniture.[62]

The emergence of this large-scale utilisation (exploitation) programme in the Forestry Department seems at first to sit awkwardly with the persistent 'deforestation degradation' narrative that pervaded forestry policy circles during the colonial era.[63] If the forest landscape was *so* degraded, how could the Department justify exploitation of the forests? The contradiction, nevertheless, was justified within the rhetoric of colonial forestry. Local populations *exploited* and *degraded* the landscape, while the Forestry Department *utilised* and *managed* it. Modern colonial forestry, apparently, could do both: exploit and protect forests in a sustainable manner. Not only this, sawmilling was even purported as being able to promote the undeniable logic of colonial forestry to the local population, and thus subsequently 'requests for reservation w[ould] be easier to obtain'.[64] The Forestry Department's exploitation programme therefore was encapsulated as helping to achieve conservation objectives and at the same time educating the 'native' population as to how to properly use their forest resources, as MacGregor argued:

> If He [the Native] sees an immediate utilisation of forest resources which bring him direct benefit in the form of timber and employment,

> the somewhat bitter pill of conservation will be more palatable. The development of sawmilling and secondary timber consuming industries will undoubtedly facilitate the introduction of the first principles of conservation.[65]

Large-scale commercial forestry, it was argued, was necessary for achieving conservation. As Jeremy Swift has observed, degradation narratives tend to be persistent, even when contradicted by empirical evidence, as they allow government actors and other elites to become stakeholders.[66] Thus, in the context of Sierra Leone, the notion that the country was rapidly being deforested provided a justification for moving the country's forest resources from local communities to a centralised department. It was not based on a nuanced understanding of material change across Sierra Leone's forests but rather represented a metanarrative that condemned local forms of forest management. Actual material change of forest cover in Sierra Leone is largely unknown due to a general lack of actual forest enumeration surveys.[67] The colonial government did not collect rigorous information of vegetation change; rather, it relied on the imaginations of colonial foresters to develop a narrative about ecological changes. European imagination was passed off as ecological fact.

Exploitation and the Private Sector

An ongoing concern for Sierra Leone and the Forestry Department in the post-World War II era was that timber production was being dominated by a Government Department, with only a few small-scale operations around the country providing an additional modest amount of timber. This changed somewhat in 1955, when Sierra Leone's first major private sawmilling enterprise, since the 1800s, was set up by the Frenchman Alfred H. Woehrling – called the Katema (and later renamed Panguma) Sawmill, which was located in Panguma, around 30 kilometres north of Kenema.[68]

Woehrling's story, and the subsequent setting up of his sawmill, was celebrated as an entrepreneurial success in Sierra Leone. As the *Sierra Leone Trade Journal* opined: 'The story of the Katema Sawmill Company is the story of how well-organised private enterprise can play an important part in the development of the country.'[69] Woehrling had moved to Sierra Leone from France in 1929 and commenced work in the gold mining sector. During World War II, he enlisted in the military, only to return to Sierra Leone to find that his mining company had ceased operations. Undeterred, he set up a sawmill in Panguma and secured a logging concession in the Dodo Hills Forest Reserve, and, using his contacts in the mining sector, he built prefabricated housing for miners employed by the large diamond mining companies. The sawmill was a great success, and Woehrling's story was recounted and celebrated in several journal and newspaper publications.[70]

Promoting private sector timber exploitation continued to be an important focus of forestry operations in the postcolonial era. The newly independent Sierra Leonean Government made its own efforts to promote in-country private commercial timber exploitation. The first major effort of this was to try and convert the Forestry Department's Utilisation Branch into a Government Company. The idea behind transformation first came about in the late 1950s. The Branch had been producing a profit for a number of years, and to help expand operations, the Colonial Government was hoping to secure funding from the Colonial Development Corporation (CDC),[71] no doubt with a mind for the financial sustainability of the industry for the postcolonial era.[72] In 1960, accountant Professor Jack (as he was named in official reports) from the CDC was sent to Sierra Leone to evaluate the potential for creating such a corporation.[73] Jack was underwhelmed by what he saw (perhaps because he was based in Nigeria, a country with a much larger forestry industry) and strongly recommended against the establishment of the corporation.[74]

Soon after independence, in 1961, the Sierra Leonean Government accepted a compromise; the Utilisation Branch was carved out of the Forestry Department and it established its own stand-alone Department, the Forestry Industries Department (FID). The two Departments were even placed under different ministries; the Forestry Department being under the Ministry of Natural Resources; while the Forest Industries Department was under the Ministry of Trade and Industry.[75] This departmental divorce appears to have been an unhappy one. The Forestry Department was undoubtedly infuriated about losing its only profitable forestry programme, left only with its forest reservation and plantation programmes. Furthermore, R.A. Price, the head of the newly established Forestry Industries Department, had a reputation for being an arrogant and difficult man to deal with.[76]

Hostilities between the two Departments erupted over the next few years. In 1963, when the Forestry Department was in need of a guesthouse for its employees, it decided to purchase prefabricated houses from Woehrling's Katema sawmill rather than from the Forest Industries Department. Price was incensed: 'The Rest House is now understood to be built by Woehrling Sawmills inside the Forestry Industries Department logging area and on a Forest Industries Department road.'[77] He pointed out to the Forestry Department that it was a 'policy of government to obtain where-ever possible all Government Department's requirements, of timber, furniture or timber pre-fabricated houses, as an economic measure, from the Forest Industries Department's factory'[78] and was not impressed with the response he got from the Forestry Department's procurement officer: 'His answer was to the effect he would place the order where he pleased.'[79] Price was victorious in the end. The Forestry Department was forced to cancel its order with Katema Sawmills and purchased the building from the Forestry Industries Department.[80]

In 1966, tensions flared up again, when the Forestry Department attempted to set up a small mill in Kasewe Forest Reserve. Price managed to get the venture blocked, as it was duplicating 'the responsibility of the Forestry Industries' and therefore, once again, contravened Government Policy.[81] The Forestry Department, and its staff, nevertheless, continued to work closely with Katema Sawmills, enjoying that it was a thorn in the side of FID and that while the Forestry Industries Department was experiencing a drop in output, production and business Katema Sawmills was booming.[82] The Forestry Department had not only lost its only profitable branch, but it had also gained an antagonistic and competitive Department.

Aside from its tension with the Forestry Department, the Forestry Industries Department largely continued the work it had previously done as a sub-departmental branch. In 1960, in an attempt to increase production, the main centre for logging operations was moved from the Kambui Hills to the Gola North Forest Reserve, the latter being particular noted for the prevalence of the commercially valuable tree species *Heritiera utilis* and *Didelotia ideae*. The FID, although turning over modest profits most years, struggled as a Department. Accessing railway wagons continued to be a problem, while electricity supply to Kenema (and hence the mill) was becoming even more erratic.[83] It had trouble securing a shop and showroom in Freetown and was annoyed by the 'unbridled importation of furniture by commercial houses'.[84] Price also voiced his annoyance when Lungi Airport (Freetown's international airport) opted to purchase £17,000 worth of furniture from the United Kingdom instead of Sierra Leone's own FID.[85] Nevertheless, FID still had some success, eventually establishing a showroom in Freetown and receiving contracts to sell furniture to embassies in Sierra Leone and Guinea.[86] FID also managed to secure an agreement with PZ and Co,[87] the largest department store in Freetown, to sell its furniture goods.[88] As part of this, there was a big promotion of its 'Made in Sierra Leone' status and how the quality of its production was equal to similar products manufactured anywhere in the world.[89] Prominent colonial forester Christopher Swabey, during his visit to Sierra Leone in 1963, also congratulated FID on its operations:[90]

> This is one of the most interesting sawmills I have seen in the tropics, using 56 species from mixed forest: apart from turning out normal industrial lumber there is a very fine furniture workshop turning out a wide range of highly sophisticated furniture of splendid quality and finish: also joinery shop turning out doors, window frames, school furniture, brooms, brushes etc: a small-wood carving shop: pre-fabricated housing units, etc.[91]

Swabey's scribbled notes on the side of his report (presumably meant only for his viewing), however, were less gratifying: he described FID's furniture designs as being 'not too revolting'.[92] Nevertheless, the public accolades he bestowed on the operations were undoubtedly a great boost. In 1965, against earlier advice, and through a Government Act,[93] FID was converted into a Government Corporation – the Forest Industries Corporation (FIC). The creation of this Corporation was initially a success. FIC shops were opened across Sierra Leone (i.e. Freetown, Koidu, Bo, Makeni); furniture was being exported to Liberia; and even West Germany was importing some of the Corporation's timber. Furthermore, exploitation was continuing on a largely sustainable yields basis in that vigorous regeneration programmes were implemented after a site had been felled for timber.[94] In 1968, it was employing around 1,000 people, making it one of the biggest industries in Sierra Leone outside of the mining sector.[95] Thus, during the late 1960s and early 1970s, a period when the Forestry Department in general was experiencing a decline,[96] FIC was experiencing a modest boom and success in its operations.

Exploitation and the Shadow State

The rise of the Siaka Stevens' patrimonial government in Sierra Leone in the 1970s and 1980s, which resulted in declining government investments in transport and industrial infrastructure, ultimately translated into an overall decline of commercial forestry in Sierra Leone. The Stevens government was initially quite keen on promoting commercial forestry in Sierra Leone. In 1975, it succeeded in setting up, with an Italian company, the Sierra Leone Timber Industries (SILETI), a timber company based in the south-east of Sierra Leone that would exploit parts of the Gola Forest. The Sierra Leonean Government was a part owner of SILETI; however, all of its operations were conducted by the Italian company. Curiously, existing forestry legislation was not used to make way for the venture; instead, a specific retrospective parliamentary act (pushed through by the Vice President) was passed to make way for SILETI's agreement:

> The Concession Agreement for Gola Forest East and West, Forest Reserves (Ratification) Act, 1977: Being an Act to Ratify and Confirm the Agreement signed on the 20th day of January, 1975 by representatives of Sierra Leone Government of the one part and the Managing Director of SILETI of the other part establishing a joint venture to exploit the Gola Forest [28 February 1975].[97]

The company was given a 25-year concession to log the Gola West and East Forest Reserves; however, within just eight years, things fell apart. The company was running at a loss, and conflict emerged between the Sierra Leonean Government and the Italian co-owner Dr Giuseppe Lamarca.[98] As a result, in 1983, the Sierra Leone Government agreed to purchase the Italian share of the company for US$1,450,000. It defaulted on this, paying only $50,000.[99] In response, Dr Lamarca succeeded in convincing the Italian Government to block $100 million of funding it had pledged for a hydroelectric dam project in the north of Sierra Leone.[100] In 1990, a settlement agreement was finally reached: the Sierra Leone Government would transfer the Sierra Leone Embassy building (worth around US$1,000,000) in Rome to Dr Lamarca, while in return Dr Lamarca would make a US$300,000 donation to the Government of Sierra Leone. This US$300,000 was allegedly transferred directly to the personal (London-based) bank account of Sierra Leone's then Foreign

Affairs Minister Dr Abdul Karim Koroma and never reached the coffers of the Sierra Leonean Government.[101] The whole debacle came to be known as the 'Dr Lamarca Affair'. The result was a massive embarrassment for the Sierra Leonean Government. During the initial establishment of SILETI, it was stated that there was a 'hope that the relationship between Sierra Leone and Italy would be further strengthened' as a result of the deal.[102] However, instead, relations soured, and the Sierra Leonean Government lost its embassy in Italy as well as funding for a much needed hydroelectric project. The unprofitable SILETI ceased to exist.

FIC did not fare much better during the same period. Despite the early rhetoric of expansion and exports in the early 1970s, by the late 1970s the Corporation was in financial dire straits. In 1978, Sierra Leone received a Le 3 million loan (around US$2.8 million) from the UK government to purchase logging trucks and a new sawmill production line. That same year, the West German government provided FIC with 30 million DM (around US$14.2 million) for the construction of a new timber complex.[103] European countries that FIC just seven years earlier had been exporting timber to were now stepping in to aid the ailing corporation. After SILETI's collapse, FIC moved its operations into Gola Forest East and West Reserves – this in addition to its existing operations in the Gola North Forest Reserve.[104] Operations in both concessions, however, came to a halt by 1989, largely due to the inability of the Corporation to operate its Kenema sawmill, which had fallen into disrepair. In 1992, the Government had to set up a small (one square mile) FIC concession in the Kambui Hills in a desperate effort to pay off debts and staff wages.[105] However, the logging of this concession never eventuated due to the escalation of the civil war conflict in Sierra Leone, which had commenced in 1991.[106] Only the privately owned Katema Sawmill thrived during the 1970s and 1980s, even after the tragic death in a road accident of its owner Mr Woehrling in 1973.[107] His son continued to manage the company, and it only formally ceased operations in the early 1990s, with the advent of the civil war.[108] Overall, with the demise of SILETI and FIC, by the early 1990s only a few small-scale logging operations, mainly run by local Lebanese families,[109] were left functioning in Sierra Leone.[110]

The overall decline of commercial forestry in Sierra Leone during the 1970s and 1980s can perhaps be best understood as a by-product of Siaka Stevens' neo-patrimonial governance regime. The governance of the Forestry Department was incorporated into the structure of the neo-patrimonial state – power was heavily invested in the Chief

Conservator of Forests,[111] while logging concessions were largely secured on an informal basis. Indeed, there was a re-amalgamation between FIC and the Forestry Department in the mid 1970s, with the Chief Conservator of Forests (i.e. Musa Feika) closely overseeing both operations.[112] Harvesting ceased to operate on a sustainable yield basis, and formal programmes of regeneration were no longer a key activity post-harvest.[113] A 1983 article in *The Ecologist* lambasted how the timber industry was destroying the country's forest, with the operations of FIC and SILETI being heavily criticised.[114] Siaka Stevens' regime also oversaw a significant retreat from rural investment, which was problematic for a timber industry that required significant inputs for the maintenance of equipment and electricity for sawmills. It also required a substantial transport infrastructure system, and, therefore, perhaps the first major setback came for FIC in 1974, when Stevens decided to dismantle Sierra Leone's railway line.[115] Subsequently, during the 1980s, the cost and time required to transport goods to and from the south-east of the country increased substantially.[116] A large-scale government-run commercial timber operation appears to have been incompatible with Sierra Leone's neo-patrimonial governance regime.

The Rise of Small-Scale Exploitation

The decline in domestic large-scale timber production during the 1980s did not, however, result in a return to high timber imports for Sierra Leone; rather, Sierra Leoneans, by and large, increasingly turned to small-scale producers to realise their timber needs. Village-level pit-sawyers around the country were producing boards, while others were harvesting tree saplings to be used as building poles – each commodity being transported around the country (generally towards Freetown) through a complex network of small-scale vehicles and traders.[117] The commercial trade in fuelwood products (charcoal and firewood) also increased during this period.[118] In response to this changing dynamic, the Forestry Department passed new Forestry Regulations in 1989 that enacted a system of fees for all trade in firewood, charcoal, poles and boards produced in the country, which would be collected at checkpoints along roads around the country. The fees were commodity based, so they varied depending on the type of commodity transported (i.e. firewood, charcoal, poles and boards), not on the timber used or the geographical location from which they derived.[119]

The outbreak of Sierra Leone's civil war in 1991 augmented this trend towards small-scale, informal, commercial forestry activities. The Forestry Department, which had already been in decline during the Siaka Stevens era, essentially ceased to operate during the civil war. During the latter part of the 1990s, its staff could hardly even conduct field trips to areas outside of Freetown; and even at times the capital itself was not safe from the ravages of the conflict. During its occupation by the rebels in 1999, Prince Palmer, then Head of the Forestry Department, was killed.[120] Indeed, state authority in Sierra Leone as a whole was falling to pieces; the relationship between the 'governing' bodies in Freetown and the marginalised rural populations was considerably disrupted.[121]

This situation opened up a critical space for reorientating the domestic forestry trade. Until 1997, Freetown was relatively insulated from the excesses of the conflict and was generally seen as a haven of safety from rebel atrocities. As a result, large numbers of rural Sierra Leoneans, driven from their homes by the war, fled to the capital and its surroundings for refuge, increasing the number of its residents as much as threefold.[122] The arrival of so many new migrants created a significant upsurge in demand for forestry products – timber for construction, and fuelwood for energy needs. Importantly, the demographic shift also provided the social means of organisation for these novel commodity chains, as the new migrants in Freetown retained close ties to their villages of origin, which served as the basis for strategic trade networks linking rural producers to urban markets.[123] This major shift in population geography was proved to be critically transformative for the forestry sector, creating foundations for new markets, commodity chain networks and livelihoods for the generally cash-poor rural majority.[124]

New small-scale forestry technologies also proliferated. For example, prior to the 1990s, chainsaws were rare in Sierra Leone and used almost exclusively by large commercial timber operations such as FIC and SILETI, and village-level timber production was modest in scale and based almost exclusively on labour-intensive pit-sawing. Following the end of the civil war, this situation began to change rapidly as a number of developments combined to stimulate the rapid proliferation of chainsaws in the country.[125] This emerged, in part, with efforts to resettle Sierra Leoneans who had been displaced during the conflict: government agencies and donors provided chainsaws to chiefdoms all over the country, training local residents in their use in order to facilitate timber production for the resettlement programme. Foreign (often illegal) timber operators, from China and neighbouring Guinea, also became more prevalent in Sierra Leone during this period, often providing chainsaw training or even chainsaws as direct payment for timber.[126] This development not only furthered the dissemination of chainsaws and training in their use for on-site Chain Saw Milling (CSM) of timber boards but also illustrated to many rural residents the profitability of the practice, which was stimulating the growth of this informal sub-sector and fostering a high demand for chainsaws in rural areas as well as contributing to the near disappearance of pit-sawing operations.

Thus the decline in domestic large-scale timber production during the 1980s did not result in a return to high reliance on timber imports for Sierra Leone – instead the formal sector was supplanted by a plethora of small-scale producers operating in off-reserve rural areas across the country, who, via complex and varied transportation and trade network, rendered the country self-sufficient in timber for the first time since the 1970s.[127] It was a substantial change, as has been observed

> the Sierra Leonean civil war and its aftermath have proved both brutally destructive and creatively productive, dismantling the status quo and generating a range of novel necessities, experiences and opportunities leading to important reconfigurations of thinking and practice with respect to forests and the forestry sector.[128]

The rapid expansion and commercialisation of the small-scale forestry trade was, however, met with consternation by the Forestry Department, which remerged as an active agency in the post-civil war era. The Department, being a colonial creation, still maintains a legislative and policy framework that continues to emphasise large-scale industrial timber production under its sole authority.[129] Thus, the emergence of informal, flexible and even (spatially) ephemeral timber supply networks represents the antithesis to the colonial model of forestry – a dramatic contrast to the centralised control of forest resources that the agency sees as its raison d'être.[130] The Forestry Department has reacted aggressively to this situation, developing prohibitively expensive commercial licensing fees and draconian fines and even jail sentences for the violation of regulations that it has designed expressly to drive small and medium-sized producers out of business.[131] Its justification of these moves blends narratives of national development and environmental degradation into a form of eco-modernisation discourse, framing small-scale producers as primitive and wasteful forest destroyers, and centralised bureaucratic forest regulation as an urgent imperative – an echo of the colonial forestry past.

Since the mid 2000s, in the post-civil war era, the Forestry Department has also made some abortive attempts to reintroduce large-scale forestry production. In 2005, the Sierra Leone government sold FIC to a Nigerian company called Gava. While FIC operations had not been in operation since the 1990s, Gava was at least able to secure the old FIC site in Kenema, and after negotiations with the Sierra Leonean Government, it was also able to secure a logging concession in the nearby Kambui Hills Forest Reserve.[132] The deal had shades of SILETI about it, as the sale of FIC and the concession was largely secured with help from the then Finance Minister of Sierra Leone, James Jonah, who was reportedly close to Gava's owner Lucky Nosakhare Igbinedion. Igbinedion was a highly controversial former governor in Nigeria who has been facing a number of corruption charges.[133] The Gava-owned FIC operations, however, have stagnated due to an ongoing bitter dispute between the Forestry Department and Gava FIC about the size of its logging concession.[134] Small-scale informal forest exploitation in Sierra Leone still reigns supreme.

Exploitation Legacy

A functioning commercial timber industry represented the ultimate objective of colonial forestry, building upon the previous programmes of forest reservation and plantations. The earlier programmes involved multiple negotiating processes to transform Sierra Leone's forests, conceptually, legally and materially into profitable spaces so that exploitation could be realised effectively. This grand vision, however, did not play out smoothly. As Chapter 2 explored, the establishment of a large-scale estate of Forest Reserves was never realised, and while, as examined in Chapter 3, plantation programmes had some success, they were overly reliant on a handful of species. The exploitation did not emerge out of carefully crafted policy but rather out of the political space afforded by the conditions of World War II. The colonial Forest Department exploitation programme was more bumbling than domineering, suffering from poor infrastructure and a problematic departmental divorce. Although there was some 'success' during the 1950s and 1960s, things ultimately fell apart.[135]

Indeed, commercial forestry, over time, did not conform to the colonial perception or to the Forestry Department's design. A lively timber production sector did emerge during the 1980s and 1990s, just not in the colonial (and now contemporary) perspective of how it should operate, but instead being conducted by a complicated mosaic of small-scale producers. A similar process has been observed in the forestry sectors in Nigeria,[136] and Ghana.[137] In the early colonial perspective, 'local communities' were imagined to have been gradually incorporated into the capitalist colonial trade system, moving from being subsistence farmers (and forest users) to being labourers in a variety of sectors – including large-scale forestry – but not the main actors within the production process, as has occurred. The decline of large-scale commercial forestry in Sierra Leone, thus, did not result in a return to the pre-commercial industry past (i.e. timber imports) and subsistence production in rural areas, but instead the small-scale operation actors have filled a space produced by the country's capitalist market.

Colonial foresters were preoccupied with getting rural Sierra Leoneans involved in producing timber products for urban and international markets – to supply goods, not to reap the profits. Instead, today, villagers across Sierra Leone are engaging with markets, supplying primary or speciality forest products to Sierra Leone's urban areas – and are usually renumerated with a small income.[138] This form of commercial forestry in Sierra Leone currently sits in tension with a Forestry Department that is trying to swim against the metaphorical current to establish a large-scale commercial forestry section and ultimately more direct control over the country's commercial forest resources. The Forestry Department's timber exploitation initiatives in Sierra Leone, however, have not just struggled with intransience from local political economic dynamics but they have also come up against the wildlife conservation movement, which has been increasingly on the rise – as the following chapter will discuss.

Notes

1. H.R. Blanford, 'Moves in the Colonial Forest Service', *The Empire Forestry Review* 26 (1947), 179; Colonial Office, *The Colonial Forest Service List* (His Majesty's Stationary Office, 1936).
2. W.D. MacGregor, 'Sierra Leone', *News Bulletin of Empire Forest Departments for the First Half of 1940* (1940), 12–13.
3. W.D. MacGregor, *Forest Department Sierra Leone: Statement of Policy* (Government Printer, 1946).

4. W.D. MacGregor, *Sierra Leone Forestry Report for the Year 1940* (Government Printer, 1941).
5. W.D. MacGregor, 'The Forest Production Programme in Sierra Leone', *Farm and Forest* 3(3) (1942), 116–17.
6. T.E. Edwardson, Sierra Leone Diaries – 6 June, 1940: RHA – MSS Afr r 210(1).
7. W.D. MacGregor, Letter to the Colonial Secretary, Freetown (11 June 1941): SLNA – A/38/41.
8. MacGregor, Letter to the Colonial Secretary (11 June 1941): SLNA – A/38/41.
9. W.D. MacGregor, Letter to the Colonial Secretary, Freetown (10 July 1941) SLNA – A/38/41.
10. W.F. Chipp, *Report on Possibilites of Development of Timber Production in Sierra Leone* (Forestry Department, 1939), 6.
11. A.H. Lloyd, 'The Operations of Military Forestry Units in France and Great Britain, 1939–44', *The Empire Forestry Journal* 23(1) (1944), 10–19.
12. W.D. MacGregor, 'Sierra Leone – A Forest Mill', *The Empire Forestry Journal* 22 (1943), 68.
13. W.D. MacGregor, *Report on the Forest Administration of Sierra Leone for the Year 1944* (Government Printer, 1945).
14. W.D. MacGregor, *Sierra Leone Annual Report of the Forestry Department for the Year 1943* (Government Printer, 1944), 1.
15. H.R. Blanford, 'UK Schools of Forestry, 1949–50', *The Empire Forestry Review* 28 (1949), 109–11; E.W. March, 'News of Members', *The Commonwealth Forestry Review* 47 (1968), 174.
16. *The Empire Forestry Journal*, H.M. The King's Birthday Honours List, 24 (1945), 2; Blanford, 'Moves in the Colonial Forest Service'.
17. W.D. MacGregor, *Sierra Leone Forestry Report for the Year 1942* (Government Printer, 1943); MacGregor, *Report on the Forest Administration of Sierra Leone for the Year 1944*.
18. R.O. Ramage, C.J. Hodgens, F.A. Baine, J. Jardine, R. Briggs, W.D. MacGregor and A.F.R. Stoddart, Minutes Of A Meeting Held In The Colonial Secretary's Office On The 6th May, 1943: TNA CO 267/679/3.
19. R.S. Pelly, *Statement by the Forest Authority, Sierra Leone Prepared for the Sixth Commonwealth Forestry Conference, 1952* (Government Printer, 1952), 10.
20. A.H. Unwin, *Report on the Forest and Forestry Problems in Sierra Leone* (Waterlow and Sons Limited, 1909).
21. A.H. Unwin, *West African Forests and Forestry* (T. Fisher Unwin Ltd, 1920);Unwin, *Report on the Forest and Forestry Problems*, 33.
22. J. Dargavel, *The Zealous Conservator: The Life of Charles Lane Poole* (University of Western Australia, 2008).
23. Unwin, *Report on the Forest and Forestry Problems.*
24. C.E. Lane-Poole, *Report on Forests of Sierra Leone* (Waterlow & Sons, 1911), 23.

25. Unwin, *Report on the Forest and Forestry Problems*,13.
26. Lane-Poole, *Report on Forests of Sierra Leone*, 30.
27. MacGregor, *Sierra Leone Annual Report of the Forestry Department for the Year 1943*, 8.
28. K. Burbridge, Communication to the colonial office - 25 June, 1927: SLNA - A/24/30.
29. Chipp, *Report on Possibilites of Development of Timber Production.*
30. M.A. Young, Communication to Lord Passfield (Government House) - 27 July, 1929: SLNA - A/43/29.
31. D.G.Thomas, 'Report: The Forest Authority, Sierra Leone', paper presented at the The British Empire Conference, London, 1924.
32. MacGregor, *Sierra Leone Annual Report of the Forestry Department for the Year 1943*, 11.
33. E. MacDonald, Letter about exploitation - Communication to the Colonial Office, 1932: SLNA - A/31/32.
34. Lane-Poole, *Report on Forests of Sierra Leone*; A. Sharpe, 'The Hinterland of Liberia', *Geographical Journal* 55(4) (1920), 299; Unwin, *Report on the Forest and Forestry Problems.*
35. E. MacDonald, Further Report by the Conservator of Forests on the classification and enumeration of timbers, 1993: TNA - FD 283/47/32.
36. E. MacDonald, Enclosure in Dispatch, Sierra Leone No. 497 - 21st December 1933: TNA - FD 283/47/32.
37. W.M. Robertson, First Half Yearly Newsletter, 1936: SLNA - E/48/30.
38. E. MacDonald, Communication to the Colonial Secretary, 19 June 1935: SLNA - A/5/45; E. MacDonald, Memorandum on Order in Council No. 19 of 1935: SLNA - A/31/34.
39. Colonial Forest Resources Development Department. *Report on the Witte Power Saw*, 1935: SLNA - A/31/32.
40. Chipp, *Report on Possibilites of Development of Timber Production*, 5.
41. W.F. Chipp, Governor's conference, 1939 Appointment of forest engineers for service in Nigeria, the Gold Coast, S/L, 1939: SLNA - A/66/39; D.G. Thomas, Communication to E. MacDonald (CoF)1936. SLNA - A/5/31; W.M. Robertson, Local Timber Exploitation Scheme, 1937: SLNA - A/15/37.
42. W.M. Robertson, *Sierra Leone Annual Report of the Forest Department for the Year 1936.* (Government Printer, 1937).
43. Robertson, *Sierra Leone Annual Report of the Forest Department for the Year 1936.*
44. Chipp, *Report on Possibilites of Development of Timber Production.*
45. Chipp, *Report on Possibilites of Development of Timber Production.*
46. H.R. Blanford, 'REVIEW: Sierra Leone: Annual Report Of The Forestry Department For The Year 1938. (Memeographed.) (Government Printer, 1939.)', *The Empire Forestry Journal* 19 (1940), 160–61.
47. Colony of Sierra Leone, *Empire Forests and the War Empire Forestry Conference, 1947.* (Government Printer, 1947). A.H. Weir, 'Sierra Leone:

Report on Forest Administration For The Year 1939', *The Empire Forestry Journal* 19 (1940), 331–32.

48. D.H. Hodgson, *Annual Report on the Forest Administration for the Year 1949* (Government Printer, 1951); W.D. MacGregor, *Report on the Forest Administration of Sierra Leone for the Year 1945* (Government Printer, 1946).
49. ITMB (Cartographer). *Map of Sierra Leone and Freetown* (International Travel Maps and Books, 2009).
50. D.P. Gamble, 'Kenema – A Growing Town in Mende Country', *The Bulletin – The Journal of the Sierra Leone Geographical Association* 7–8 (1964), 9–12; D.P. Gamble, 'Urbanisation in West Africa', *The Bulletin – The Journal of the Sierra Leone Geographical Association* 7–8, (1964), 7–8.
51. *Daily Mail*, 'Forest Reserves Protect Crops and Soil', December 30, 1959, 1; *Sierra Leone Trade Journal*, 'Sierra Leone Forest Industries Corporation', 8(2) (1968), 38–41.
52. A.K. Nicol, *Annual Report on the Forest Administration of Sierra Leone for the Year 1952.* (Government Printer, 1953); R.S. Pelly, *Annual Report on the Forest Administration for the Year 1953* (Government Printer, 1955); W.M. Robertson, 'REVIEW: Second Annual Report of The Forestry Adviser, 1951', *The Empire Forestry Review* 32(1953), 374–75.
53. A.F.A. Lamb, *1956 Report on the Forest Administration* (Government Printer, 1957); A.F.A. Lamb, *Report on the Forest Administration of Sierra Leone 1957* (Government Printer, 1959); A.F.A. Lamb, *Annual Report on the Forest Administration of Sierra Leone 1958/1959* (Government Printer, 1960); Nicol, *Annual Report on the Forest Administration of Sierra Leone for the Year 1952*; A.K. Nicol, *Report on the Forest Administration of Sierra Leone for the Year 1955* (Government Printer, 1956).
54. Lamb, *1956 Report on the Forest Administration*; Nicol, *Annual Report on the Forest Administration of Sierra Leone for the Year 1952*; A.K. Nicol, *Report on the Forest Administration of Sierra Leone for the Year 1954* (Government Printer, 1955); Pelly, *Statement by the Forest Authority, Sierra Leone Prepared for the Sixth Commonwealth Forestry Conference, 1952* (Government Printer, 1952).
55. Nicol, *Annual Report on the Forest Administration of Sierra Leone for the Year 1952*; Nicol, *Report on the Forest Administration of Sierra Leone for the Year 1954*; R.S. Pelly, *Annual Report on the Forest Administration for the Year 1951* (Government Printer, 1952); D.H. Hodgson, *Report on the Forest Administration of Sierra Leone for the Year 1948* (Government Printer, 1949); Hodgson, *Annual Report on the Forest Administration for the Year 1949*; D. H. Hodgson, *Annual Report of the Forest Department for the Year 1950* (Government Printer, 1952).
56. R. Best, *A History of the Sierra Leone Railway 1899–1949* (Freetown, 1949).
57. A later Sierra Leone Forestry Head even wrote a novel about the illicit mining trade – *The Mocking Stones* – based upon his experience of the trade while

working for the Forestry Department. See P.D. Palmer, *The Mocking Stones* (Longman, 1982).

58. Nicol, *Report on the Forest Administration of Sierra Leone for the Year 1954.*
59. R.S. Pelly, 'Shingles in Sierra Leone', *Farm and Forest* 8(1) (1947), 18–19; M.H. Dorman, *The Review of Government Departments During 1959*, Sessional Paper No.1 of 1960 (Government Printer, 1960).
60. *Daily Mail*, 'Full Scale Works in Kenema Mills', November 26, 1957, 2.
61. I.G. Bulmer, *Report on the Forest Administration of Sierra Leone 1959/1960* (Government Printer, 1960).
62. *Daily Mail*, 'Furniture for Sierra Leone's Hotel is Being Made Locally', 1960, 3.
63. J. Fairhead and M. Leach, *Reframing Deforestation: Global Analyses and Local Realities with Studies in West Africa* (Routledge, 1998).
64. MacGregor, *Sierra Leone Annual Report of the Forestry Department for the Year 1943.*
65. MacGregor, *Forest Department Sierra Leone.*
66. J. Swift, 'Desertification: Narratives, Winners & Losers', in M. Leach and R. Mearns (eds), *The Lie of the Land: Challenging Recived Wisdom on the African Envrionment* (James Curry, 1996), 73–90.
67. R.A. Wadsworth and A.R. Lebbie, 'What Happened to the Forests of Sierra Leone?', *Land* 8(5) (2019), 80.
68. Nicol, *Report on the Forest Administration of Sierra Leone for the Year 1955*; J.S. Sawyerr, *Ten-Year Plan of Forestry Development for Sierra Leone: 1964/5 - 1973/1974* (Ministry of Natural Resources, 1964), 8.
69. *Sierra Leone Trade Journal*, 'Private Enterprise Flourishes in Sierra Leone' 1 (1961), 40.
70. *Daily Mail*, 'The Story of a Sawmill in Kenema District', April 20, 1960, 8; *Sierra Leone Trade Journal*, 'Private Enterprise Flourishes', 40–41; *Sierra Leone Trade Journal*, 'Private Enterprise Flourishes in Sierra Leone', 4 (1963), 80–81.
71. The Colonial Development Corporation was established as a statutory corporation in 1948 by the British Government to assist its colonies in the development of agriculture and related activities. Following the independence of many colonies, it was renamed the Commonwealth Development Corporation in 1963 (see M. Cowen, 'Early Years of the Colonial Development Corporation: British State Enterprise Overseas during Late Colonialism', *African Affairs* 83(330) (1984), 63–75; E.R. Wicker, 'The Colonial Development Corporation (1948–54)', *The Review of Economic Studies* 23(3) (1955), 213–28.
72. *Daily Mail*, 'Expert to Review Future of Forestry Industry', July 17, 1958, 6; *Daily Mail*, 'Forestry Industry May Be Commercialised Soon', July 5, 1958, 1; *The Empire Forestry Review*, 'Sierra Leone Forest Industry', 37 (1958), 398.
73. Lamb, *Annual Report on the Forest Administration of Sierra Leone 1958/1959.*
74. Lamb, *Annual Report on the Forest Administration of Sierra Leone 1958/1959.*
75. R.A. Price, Communication to the Permanent Secretary of the Ministry of Trade and Industry, 17th February 1966: SLNA D/10/7.

76. John Wools, personal communication, 2011; Peter Savill, personal communication, 2011.
77. R.A. Price, *Forestry Industries Department Monthly Progress Report for the Month of September 1963* (Forest Industries Department, 1963), 1.
78. Price, *Forestry Industries Department Monthly Progress Report for the Month of September 1963.*
79. Price, *Forestry Industries Department Monthly Progress Report for the Month of September 1963.*
80. R.A. Price, *Forestry Industries Department Monthly Progress Report for the Month of October 1963* (Forest Industries Branch, 1963).
81. H.S. Gobeer, Communication to MTI and MANR – 'Forest Industries Corporation', 8 March 1966: SLNA D/10/7; Price, Communication to the Permanent Secretary of the Ministry of Trade and Industry, 17th February 1966: SLNA D/10/7.
82. J.S. Sawyerr, *Report on the Forest Administration of Sierra Leone for the year April, 1963 to March, 1964* (Government Printer, 1965); Peter Savill, personal communication 2011; John Wools, personal communication, 2011.
83. C. Curling, *Forestry Industries Department Monthly Progress Report for the Month of February 1963* (Forest Industries Department, 1963); R.A. Price, *Forest Industries Montly Progress Report for the Month of Nov, 1962* (Forest Industries Department, 1962); R.A. Price, *Forestry Industries Department Monthly Progress Report for the Month of April 1964* (Forestry Industries Department, 1964).
84. Minister of Trade and Industry, *Sale of Forest Industries Furniture on Deferred Payments Basis* (Ministry of Trade and Industry, 1964); Price, *Forestry Industries Department Monthly Progress Report for the Month of October 1963*; R.A. Price, *Forestry Industries Department Monthly Progress Report for the Month of May 1963* (Kenema: Forest Industries Department, 1963); R.A. Price, *Forestry Industries Department Monthly Progress Report for the Month of August 1963* (Forest Industries Department, 1963).
85. R.A. Price, *Forestry Industries Department Monthly Progress Report for the Month of June 1963* (Forest Industries Department, 1963).
86. Curling, *Forestry Industries Department Monthly Progress Report for the Month of February 1963.*
87. PZ is short for Paterson Zonchonis – a trading company that was set up in Sierra Leone in 1879 by George Paterson and George Zochonis. Now known as PZ Cussons, it no longer operates in Sierra Leone but is now one of the largest companies in the world (see P. Zochonis, *Paterson Zochonis Plc: A Century of Enterprise* (The Company, 1984).
88. Price, *Forestry Industries Department Monthly Progress Report for the Month of April 1964.*
89. Forest Industries Corporation, *Sierra Leone Forest Industries Corporation, Annual Report and Accounts for the Year Ended 31st December 1966*

(BrownKnight and Truscott Ltd, 1967); *Sierra Leone Trade Journal*, 'Furniture from Kenema Compares Favourably Anywhere', 5(1) (1965), 6–8.
90. C. Swabey, Jottings From Forestry Adviser's Tour Diaries, 1963: RHA MSS Brit Emp s. 459.
91. Swabey, Jottings From Forestry Adviser's Tour Diaries, 1963: RHA MSS Brit Emp s. 459.
92. Swabey, Jottings From Forestry Adviser's Tour Diaries, 1963: RHA MSS Brit Emp s. 459.
93. *Forest Industries Corporation Act, 1964.*
94. J.E.D. Fox, 'Exploitation of the Gola Forest', *Journal of West African Science Association* 13(2) (1968), 185–210; M. Illes, P.S. Savill, and G. Koker, *Gola Forest Reserves Sierra Leone, Interim Management Plan* (RSPB, 1993); *Sierra Leone Trade Journal*, 'Furniture to be Exported to Liberia', 8(1) (1968), 20, 22; *Sierra Leone Trade Journal*, 'Sierra Leone Forest Industries Corporation', 38–41.
95. *Sierra Leone Trade Journal*, 'Furniture to be Exported to Liberia', 20, 22.
96. Peter Savill, personal communication, 2011.
97. *The Concession Agreement for Gola Forest East and West, Forest Reserves (Ratification) Act, 1977*, 1.
98. D. Tuboku-Metzger, 'Forest Exploitation in Sierra Leone: A Tale of Devastation', *The Ecologist* 13(6) (1983), 241.
99. J. Alie, 'History Alive! (Final) A Reaction to Ernest Koroma's Recent Policy Statement', *Awareness Times*, August 8, 2007; *Sierra Herald*, 'The Government of Sierra Leone's Indebtedness to Dr Lamarca', 9(1) (2011).
100. J. Alie, 'History Alive!'.
101. *Sierra Herald*, 'The Government Of Sierra Leone's Indebtedness to Dr Lamarca'; Alie, 'History Alive!'; S.O. Blyden, 'Dr. Sama Banya: Deaf or Fen Plabar (Quarrelsome)?', *Awareness Times*, December 6, 2011.
102. *Sierra Leone Trade Journal*, 'Gola Forest to be Exploited', 13(4) (1973),110.
103. Office of the President, *Sierra Leone: 12 Years of Economic Achievement and Political Consolidation under the APC and Dr Siaka Stevens: 1968–1980* (State House, 1980); Tuboku-Metzger, 'Forest Exploitation in Sierra Leone', 239–41.
104. A.G. Davies, *The Gola Forest Reserves, Sierra Leone: Wildlife Conservation and Forest Management* (IUCN, 1987); Illes et al., *Gola Forest Reserves Sierra Leone*.
105. Illes et al., *Gola Forest Reserves Sierra Leone*.
106. Illes et al., *Gola Forest Reserves Sierra Leone*.
107. M.B. Feika, *Report on the Forest Administration of Sierra Leone 1963/1964 to 1973/74* (Government Printer, 1975).
108. Peter Savill, personal communication, 2011.
109. Lebanese immigrants first came to Sierra Leone in the 1890s after a silk-worm crisis struck their homeland, then part of the Ottoman Empire. Since this time, they have established themselves as a merchant class in Sierra

Leone involved in a wide array of different economic sectors (see H.L. van der Laan, *The Lebanese Traders in Sierra Leone, Change and Continuity in Africa* (Mouton, 1975).

110. D. Hartley, 'Forest Resource Use and Subsistence in Sierra Leone', PhD (UCL, 1992); M. Leach, 'Images of Propriety: The Reciprocal Constitution of Gender and Resource Use in the Life of a Sierra Leonean Forest Villages', PhD (University of London, 1990); M. Leach, *Rainforest Relations: Gender and Resource Use among the Mende of Gola, Sierra Leone* (Edinburgh University Press, 1994); P.G. Munro and G. van der Horst, 'Breaks with the Past: Conflict, Displacement, Resettlement and the Evolution of Forest Socioecologies in Sierra Leone', in J. Lahai and T. Lyons (eds), *African Frontiers: Insurgency, Governance and Peacebuilding in Post-Colonial States* (Ashgate Publishing, 2015), Chapter 9: 119–30.
111. A. Grainger and W. Konteh, 'Autonomy, Ambiguity and Symbolism in African Politics: The Development of Forest Policy in Sierra Leone', *Land Use Policy* 24(1) (2007), 42–61.
112. Grainger and Konteh, 'Autonomy, Ambiguity and Symbolism', 42–61; W. Konteh, 'Forest Resource Management in Sierra Leone: A Critique of Policy Formation and Implementation', PhD (University of Leeds, Leeds, 1997).
113. Tuboku-Metzger, 'Forest Exploitation in Sierra Leone', 239–41; Peter Savill, personal communication, 2011.
114. Tuboku-Metzger, 'Forest Exploitation in Sierra Leone', 239–41.
115. The railway was poorly built; therefore, this in part justifies its dismantling. However, it ran through the south-east of Sierra Leone where Sierra Leone's Mende ethnic group was dominant: this group tended to support the SLPP, thus its dismantling was linked to weakened transport links in a region that tended to oppose Steven's rule (See Best, *A History of the Sierra Leone Railway 1899–1949* (Freetown, 1949); M.C. Ferme, *The Underneath of Things: Violence, History, and the Everyday in Sierra Leone* (University of California Press, 2001).
116. A. Jalloh, 'The Fula and the Motor Transport Business in Freetown, Sierra Leone', *African Economic History* 26 (1998), 63–81; J.B. Riddell, 'Internal and External Forces Acting upon Disparities in Sierra Leone', *Journal of Modern African Studies* 23(3) (1985), 389–406.
117. P.G. Munro and G. van der Horst, *The Domestic Trade in Timber and Fuelwood Products in Sierra Leone: Current Dynamics and Issues* (FAO/EU 2012).
118. E.K. Alieu, 'Building on Local Foundations: Enhancing Local Community Support for Conservation', *Unasylva* 61 (2010), 22–27; O.R. Davidson, *Energy Use Patterns, Sierra Leone* (International Development Research Centre, 1985); J. Kamara, *Firewood Energy in Sierra Leone – Production, Marketing, and Household Use Patterns* (Verlag Weltarchiv, 1986); Munro and Van der Horst, *The Domestic Trade in Timber and Fuelwood Products.*
119. *The Forest Regulations, 1989 (1990).*

120. Munro and Van der Horst, 'Breaks with the Past', 119–30.
121. P.G. Munro, 'Deforestation: Constructing Problems and Solutions on Sierra Leone's Freetown Peninsula', *The Journal of Political Ecology* 16 (2009), 104–24.
122. P.G. Munro and G. Hiemstra-van der Horst, 'Conserving Exploitation? A Political Ecology of Forestry Management in Sierra Leone', *The Australasian Review of African Studies* 32(1) (2011), 59–72.
123. P.G. Munro, G. van der Horst and S. Healy, 'Energy Justice for All? Rethinking Sustainable Development Goal 7 through Struggles over Traditional Energy Practices in Sierra Leone', *Energy Policy* 105 (2017), 635–41.
124. Munro and Van der Horst, 'Breaks with the Past', 119–30.
125. Munro and Van der Horst, *The Domestic Trade in Timber and Fuelwood Products*; P.G. Munro and G. van der Horst, *The Governance and Trade of Wood-Based Products in and Around the Kambui Hills North Forest Reserve* (PAGE, 2012).
126. G. Hiemstra-van der Horst, '"We Are Scared to Say No": Facing Foreign Timber Companies in Sierra Leone's Community Woodlands', *The Journal of Development Studies* 47(4) (2011), 574–94; G. Hiemstra-van der Horst, P.G. Munro and S.P.J. Batterbury, 'Les réseaux illégaux du pillage: La demande globale de bois et la (re)commercialisation des forêts d'Afrique de l'Ouest', *Écologie & Politique* 42 (2011), 47–58; Munro and Van der Horst, *The Domestic Trade in Timber and Fuelwood Products*.
127. Munro and Van der Horst, *The Domestic Trade in Timber and Fuelwood Products*.
128. Munro and Van der Horst, 'Breaks with the Past', Chapter 9: 125.
129. Munro and Hiemstra-van der Horst, 'Conserving Exploitation?', 59–72.
130. Munro and Van der Horst, 'Breaks with the Past', Chapter 9, 119–30.
131. Hiemstra-van der Horst, '"We Are Scared to Say No"', 574–94; Hiemstra-van der Horst et al., 'Les réseaux illégaux du pillage', 47–58.
132. Munro and Van der Horst, *The Governance and Trade of Wood-Based Products*.
133. B. Akinola, *Authority Stealing: How Greedy Politicians and Corporate Executives Loot the World's Most Populous Black Nation* (Authorhouse, 2012); D. Imoudou, 'The Real Story: The Re-Trial of Ex-Governor Lucky Igbinedion', *Elendu Reports*, 23 February, 2011; M. Vandi, 'Sierra Leone: On Government's Ban on Timber Logging and Exportation', *Concord Times*, 28 February 2008.
134. B.M. Kamara, Letter to District Forestry Office 'Timber license to Gava Forest Industries Co-operation [sic]' (NaCEF, 2005); M.A. Bockarie, *Summary of Trees Felled by the Gava Forest Industries* (Forestry Division, 2007).
135. P. Von Hellermann, *Things Fall Apart?: The Political Ecology of Forest Governance in Southern Nigeria* (Berghahn Books, 2013).
136. R.A. Cline-Cole, 'Dryland Forestry: Manufacturing Forests & Farming Trees in Nigeria', in M. Leach and R. Mearns (eds), *The Lie of the Land: Challenging Received Wisdom on the African Environment* (James Currey, 1996), 122–39.

137. K.S. Amanor, 'Farmers, Forestry and Fractured Environmentalism in Ghana's Forest Zones', in R.A. Cline-Cole and C. Madge (eds), *Contesting Forestry in West Africa* (Ashgate, 2000), 307–21.
138. J.C. Ribot, 'Forestry and Democratic Decentralization in Sub-Saharan Africa: A Rough Review', in L.A. German, A. Karsenty and A. Tiani (eds), *Governing Africa's Forests in a Globalized World* (Earthscan, 2011), 29–55.

CHAPTER 5

Wildlife Conservation

In 1933, the Society for the Preservation of the Fauna of the Empire (now known as Fauna and Flora International) sent an emissary to Sierra Leone to evaluate the state of the colony's wildlife. The Society, in conjunction with European governments,[1] was in the process of establishing an International Convention for the Protection of Flora and Fauna, and as such the emissary's visit was about informing the upcoming legislation as well as providing recommendations for Sierra Leone's wildlife conservation programme. The emissary sent to Sierra Leone, Colonel A.H.W. Haywood, made no secret of the Society's objective:

> If a reasonable proportion of the fauna of British West Africa is to be conserved it is imperative that action be taken at once, otherwise there will be nothing left to preserve and future generations will rightly hold us responsible.[2]

Haywood's expedition to Sierra Leone, while politely received, gained little favour from the colony's Colonial Government. District Commissioners around the colony reported to Haywood that they felt that wildlife stock in the country was in a healthy state, arguing that the current *1901 Wildlife Ordinance* was sufficient legislation for future conservation efforts and that while the idea of Game Reserves was fine they were not practical for the Colony at this stage:

> I consider the existing Game Ordinance [*1901 Wildlife Ordinance*] is sufficiently effective and in my opinion, there is no necessity for

> any amendments … My experience is limited to seven years, but my impression is that there is no decrease in the wild life.[3]
>
> From my personal experience of five years in the Northern Province. I have no reason to think that wild life is decreasing. [4]
>
> I do not think that elephants require further protection than that already afforded to them by the Game Regulations … Game sanctuaries are of course desirable, but I do not think they are practicable or really necessary in Sierra Leone.[5]

Despite these tepid responses, after his visit to Sierra Leone, Haywood proceeded to offer a number of recommendations for the colony, which included the suggestion that wildlife regulation should be placed under the jurisdiction of the Forestry Department and that, if possible, all Forest Reserves should be converted into Game Reserves. He noted, in particular, that there was an urgent need to convert the Gola Forest Reserves into either a Game Reserve or a National Park to ensure the long-term protection of its 'Pigmy Elephants'.[6] Colonel Haywood's recommendations came with some weight as the Society's membership was populated by European elites and as such it had considerable influence over colonial affairs during the period.[7] Nevertheless, none of Haywood's recommendations were implemented; instead, only a couple of minor pieces of legislation were passed to restrict the hunting of one animal – the banded duiker (*Cephalophus zebra*) – along with some more minor restrictions on elephant hunting.[8]

Confidential communication within the Sierra Leone colony at the time provides some insights into why there was such a weak response to Haywood's recommendations. In particular, the colony's Governor, Arnold Hodson, played a fundamental role in undermining the Society for the Preservation of the Fauna of the Empire's activities. Hodson was Sierra Leone's Governor between 1931 and 1935 (see Figure 5.1). He was a career colonialist who had previously served the Empire in Australia, South Africa, Somaliland (Somalia), Southern Abyssinia (Ethiopia) and the Falklands. After his time in Sierra Leone, he served as Governor of the Gold Coast (Ghana) until 1941. Like Haywood, Hodson had a great interest in the 'pygmy elephants' of the Gola Forest; however, he was more interested in hunting them than conserving them. He had been an avid hunter throughout his colonial career and had even published a couple of books about his hunting expeditions during colonial service.[9]

Hodson's first visit to the Gola forest was in 1932, a year before Haywood's visit to the Colony on the Society's behalf. The visit was

Figure 5.1 Portrait of Sir Arnold Wienholt Hodson held by the National Portrait Gallery. The painting was done by Bassano in 1928. © National Portrait Gallery, London.

supposed to be a tour of the Liberian frontier; however, Hodson took the 'opportunity of visiting the Gola Forest, the greater part of which is still unknown to Europeans' and proceeded to hunt elephants.[10] Hodson became convinced that three species of elephant existed in the forest – they were called (in decreasing size), in Mende, the *Baigbwowei*; The *Jangai* and the *Sumbi* (pygmy elephant). After a 'hard day's hunting' he secured a specimen (of what he believed to be) the *Jangai* species and noted that in future he would 'leave no stone unturned to try and secure, in due course, a Sumbi [pygmy elephant] skull'.[11] Hodson's elephant-hunting visit turned into an obsession, and upon his return to Freetown he immediately sent communication dispatches to London: to the British Museum, the Natural History Museum and the Colonial Office, recounting his hunting expedition. He contended that his discovery of new elephant species would be of 'great interest [for] scientists of all countries',[12] and he organised for the head of the *Jangai* specimen to be shipped to the Natural History Museum. He then sent a request to the Colonial Office asking them to fund a taxidermist's expedition to Sierra Leone to assist with the procurement of a pygmy elephant specimen.[13] He also suggested that the Colonial Office should publish his 'findings' in *The Field* – a widely circulated sports hunting magazine.[14] Hodson evidently wanted to attract European Sport hunters to Sierra Leone, and while elephants were technically a protected animal in the Gola Forest Reserves, the law allowed for the Governor (i.e. Hodson) to be able to provide anyone with special permission for hunting. He personally used this gubernatorial privilege over the next three years (1932–1935) to 'collect' eight other elephant skulls from the Gola Forest, from the three supposedly different species of elephant.[15]

The staff at the Museums and the Colonial Office in London ridiculed Hodson's requests in confidential communications:

> This is a rather embarrassing query, as it is difficult to resist the impression that the collection of elephant skulls for the Museum was more in the nature of a hobby of the Governor than an official matter.[16]
>
> It will be a little unfortunate, therefore, if we choose just this time to send to the 'Field' what I take the Governor means to be an incitement and encouragement to big game hunters to go out to Sierra Leone and slay as many pigmy elephants and perhaps other elephants as they can manage. … I believe it is generally admitted that big game hunting is one of the worst ways by which a Colonial Government can bring money into the country, but although I think it quite likely that big game hunters, while not afraid of an elephant, would shrink before the onslaught of anopheles or the even more dreaded yellow fever mosquito (whose name I forget), still I cannot help wishing the Governor had not made this request. Why on earth couldn't he have sent it to the 'Field' himself, as he did the letter the other day about the big tarpon [snake] which he or one of his staff caught. … South Kensington [The Natural History Museum] is another matter,

> though personally, I am always sceptical about the advantage of adding another stuffed animal to that extremely overcrowded and ill ventilated building. Still, as the Governor says, 'it is, of course, impossible to guarantee that one of these animals will be shot, for in all sport, especially in elephant hunting, there is always an element of uncertainty'. So that it is quite on the cards that the taxidermist himself might be the hunted instead of the hunter.[17]

No taxidermist was ever sent to Sierra Leone, and while there was some initial optimism about the elephant specimens that Hodson sent to London,[18] a closer examination revealed that there were no smaller species of elephant in Sierra Leone, rather the smaller specimens came from immature forest elephants (*Loxodonta African cyclotis*), as Guy Dollman, an elephant expert, concluded in 1935: 'The Gola Forest tusks undoubtedly belong to a young animal … [and] there can be very little doubt that all the "dwarf" [i.e. pygmy] elephant ivory which has been sent home from West and Central Africa during the last 20 years is that of young or immature animals.'[19] Hodson was hunting baby elephants, not pygmy ones.

The period that Hodson was keenly hunting 'pygmy elephants' in the Gola Forest (1932–1935) coincided with Haywood's visit (in 1933) and the recommendation to convert the Gola Forest Reserves into a National Park. It is therefore unsurprising that Hodson, in internal confidential communication, expressed his direct opposition to the National Park's creation:

> Colonel Haywood also mentioned specially part of the Gola Forest [to be converted into a National Park] because it shelters pigmy elephants, bongo, chimpanzees, colobus and banded duiker. It also holds pigmy hippopotamus. But these animals … are believed to be local to that area in Sierra Leone, are not really rare … as they are not decreasing, so far as can be learned, they cannot be said to be inadequately protected. The pigmy elephant is wholly protected within the Forest Reserve … There is therefore no need at present in this Protectorate for a National Park or Game Reserve or reserve of inadequately protected species.[20]

Hodson, however, surmised that likely there would be intense 'pressure of European Opinion' on the Sierra Leonean Colonial Government 'to spend some money on th[e] matter of preservation and to create some sanctuary'. As such, he conceded that perhaps some part of the Gola forest *could* become a Game Reserve; however, it should be 'alterable and controllable, and not a permanent setting aside of the area as a National Park'.[21] Hodson's speculation was salient, and the *International Convention for the Protection of Fauna and Flora*, which was published later that year (1933), included a requirement that all Colonies must establish areas designated to protect wildlife (Article 7).[22] In response, therefore, the Sierra Leonean Colonial Government, in a communication with the Colonial Office in London, agreed to set aside the Gola East and Gola North Forest Reserves as Game Reserves,[23] although, in a later communication, Hodson reiterated that when it was gazetted it should *not* be considered a 'strict nature reserve' in the conventional definition.[24] Hodson wanted a 'symbolic' Game Reserve to appease the Society's demands, not a formal one.

The proposed Gola Game Reserve, however, was never created, and it seems likely this was once again due to Hodson's interventions. The politically connected Hodson (he was knighted in 1934) was somehow able to get himself assigned as a plenipotentiary to oversee the implementation of the *International Convention for the Protection of Fauna and Flora*. He was one of five representatives from the British Government, and the only one based in West Africa.[25] No National Parks were ever created in West Africa during Hodson's tenure on the plenipotentiary.

Thus, one European individual, Arnold Hodson, was able to have great influence over the wildlife conservation trajectory of Sierra Leone. Hodson, however, is not a lone figure in having had a prominent role in shaping wildlife outcomes, and as the rest of the chapter explores, individuals – mainly European foreigners – feature prominently in setting up the foundations for wildlife conservation in Sierra Leone.

Wildlife and Forests in Africa

Out of the four Sierra Leonean 'forest' programmes explored this book, wildlife conservation is the most distinct. Initially, it was not a priority for the Forestry Department, and emerging concerns and policies surrounding wildlife in Sierra Leone during the colonial era, by and large, occurred outside the confines of the Department. It was not until 1967, in the postcolonial era, that a Wildlife Branch was established under the jurisdiction of the Forestry Department. Wildlife conservation, nevertheless, represents an important piece of the puzzle in Sierra Leone's forest history, and wildlife conservation in Sierra Leone has co-opted much of the previous colonial institutional and legal forestry infrastructure to realise its objectives.

Sierra Leone presents a contrasting example to many of the existing environmental histories of wildlife in Africa. Current African environmental histories on wildlife in Africa, notably by Jane Carruthers, Nancy Jacobs and Roderick Neumann,[26] have focused predominantly on nations in southern and eastern Africa, with environmental histories of wildlife in West African nations effectively non-existent. Paradoxically, this lacuna is due to a range of environmental and historical reasons.

In coastal West Africa, in comparison to the rest of the continent, questions of forests *and* wildlife are much more entangled – institutionally and materially – as it is home to large extensions of tropical rainforest. Southern and eastern Africa contain more open-plain savannah landscapes and are home to much more extensive populations of large fauna (e.g. elephants, giraffe, buffalo). The greater abundance of large animals are easier to spot due to the open nature of the landscape, and thus wildlife related-economies, both for colonial big game hunters and contemporary tourists alike,[27] have been much more substantial in southern and eastern Africa.[28] Many nations in eastern and southern Africa were also settler colony nations,[29] rather than the indirect governance approach adopted in much of West Africa, which has also provided locally based colonial hunter and tourism economies. As such, wildlife governance in these nations has often fallen to a distinct government department. [30] In coastal West Africa, in contrast, wildlife economies are limited, and wildlife is generally found in forested areas – hence, large-scale wildlife management emerged later on in West Africa and has tended to be a subsection of Forestry Departments. In Ghana, for example, The Wildlife Division is situated within the Forestry Commission of Ghana; in Nigeria, wildlife management is overseen by the Federal Department of Forestry;[31] and in Sierra Leone, as detailed below, wildlife conservation would become a pillar of the Forest Department in its postcolonial era.

Wildlife Conservation

The Sierra Leonean Colonial Government's first formal attempt to regulate the protection of wildlife in the colony appears to have occurred at the beginning of the twentieth century with the passing of the *Ordinance for the Preservation of Wild Animals, Birds and Fish, 1901*. The impetus for passing the law, however, did not come from within Sierra Leone but was rather the result of a Colonial-Office drafted piece of legislation designed for all British colonies in Africa and was informed by the colonial experience in eastern and southern Africa.[32] Indicative of this is that the Sierra Leonean legislation included protection for animals that available evidence indicates have never existed in Sierra Leone, such as zebras, rhinoceroses, gorillas and giraffes, among others.[33] The core objective of the legislation was to regulate European sport hunting; however, no specific body was created in Sierra Leone to oversee the legislation, and its enforcement was largely left to District Officers[34] around the colony. Overall, sports hunting in Sierra Leone was rarely conducted by visiting European enthusiasts[35] but rather by employees within Sierra Leone's colonial service,[36] including numerous staff members of the Forestry Department.[37] Arguably, the most prominent and avid colonial hunter in Sierra Leone was W.B. Stanley, who served as the Northern Commissioner in Sierra Leone during the first three decades of the 1900s, and he became known as an early expert on wildlife in Sierra Leone, publishing numerous books and articles on the topic.[38]

There was an ongoing debate during this early colonial period about the need to establish Game Reserves or National Parks around Sierra Leone to ensure the future protection of the colony's animals. In 1909, the visiting forester Arthur Unwin was perhaps the first colonial officer to argue in favour of the establishment of a Game Reserve in Sierra Leone after he visited the Gola Forest in the south-east of the country. He suggested that much of the Gola Forest should be made into a Game Reserve to help protect the area's elephant population. His vision was grandiose and included an elephant-hunting zone for Europeans on the edge of the proposed reserve and an elephant-taming scheme to help allay the colony's transportation woes.[39] The proposal was never implemented. The next proposition came from Lionel Palfreman, who headed the Forestry Department between 1917 and 1921. During his time in Sierra Leone, Palfreman formally proposed that the Loma Mountain Forests, in the north of Sierra Leone, should be declared a Game Reserve to help protect the elephants, hippopotami, bush cows and other wild animals in the area.[40] Palfreman's proposal, while receiving some support, was ultimately dismissed by the nearby District Commissioner in Kabala and the Colonial Government as not being feasible or urgently needed at the time:[41]

> I do not think the game of this country is in any immediate danger of extinction and though it is certainly well to prepare beforehand for such an eventuality, yet I hardly think it is sufficiently near to justify the necessary increase of staff and expenditure at present.[42]
>
> If a reserve of this nature is to be of any value it will need rangers. In view of the DC [District Commissioner], … and of the financial state of the colony, I think this should stand over for the present. I concur with the DC that there is no immediate danger of the game becoming extinct.[43]

Actions to protect wildlife, it would seem, could only be justified if it was perceived to be in immediate danger. Thus, during the first three decades of the twentieth century, colonial officials ultimately surmised that Sierra Leone's wildlife population was in a healthy state.

Stanley, the colony's resident wildlife expert, argued that wildlife numbers were healthy due to the limited number of European sport hunters as well as an overall lack of arms in the colony. Since the 1898 Hut Tax War, the Colonial Government had restricted the spread of arms among the native population, and 'hunters' represented only a small professional class.[44] Stanley also had a different reading of the landscape to forestry officials; he noted that around Sierra Leone 'natural cover of every description is abundant'.[45] Deforestation crisis narratives, it would seem, were largely the purview of colonial foresters during the early part of the twentieth century. Some hunting laws, however, were still passed. For example, in 1929, 'dazzling lights', used for night-time hunting, were banned, while in 1924, the Forestry Ordinance was amended to make all Forest Reserves in Sierra Leone de facto reserves for elephants. Subsequently, the hunting of elephants in these reserves could only legally be done with the 'special permission of the Governor' [46] – the Governor, between 1931 and 1935, being Arnold Hodson of pygmy elephant hunter fame.

Commercial Wildlife

Despite pressure from the Society for the Preservation of the Fauna of the Empire during the 1930s, wildlife conservation ultimately initially gained little traction in Sierra Leone. Concerns about wildlife did arise again after World War II (1939–1945); however, the policy focus was not on its preservation but rather on its eradication. Monkeys, and their propensity to ravage crops, had been a perennial problem for the colonial government in Sierra Leone.[47] After World War II, the monkey population was reported to have dramatically increased, presenting a plague-like issue for farming across the country. The increase was attributed to two factors. First, the country's leopard population, a natural predator of monkeys, was decimated during World War II by hunters who were selling their skins to British servicemen based in Freetown at the time, affecting the 'natural' balance. Second, the increasing influence of Islam and Christianity had meant that the consumption of monkey bushmeat was becoming increasingly taboo among major segments of the rural population, which was in contrast to neighbouring Guinea and Liberia, where there was a thriving bushmeat market.[48]

The Government's response to the issue was extreme: monkey drives were organised to try and rid farming regions of the pests. Rewards were paid for each monkey head brought into colonial agricultural centres, and between 1947 and 1962 nearly a quarter of a million monkeys were reportedly killed as a part of the programme.[49] From around 1960, Liberian hunters began to operate in conjunction with those involved in the monkey drives, taking the monkey bushmeat (only the heads were needed to be taken to the agricultural centres) across the border to be sold in Liberia's apparently thriving bushmeat markets. From this time onwards, the presence of Liberian hunters sourcing bushmeat in eastern Sierra Leone appears to have been a common phenomenon and was later seen as a major issue for primate conservation in Sierra Leone.[50]

Primates were not just exported in bushmeat form but also, throughout most of the twentieth century, there was a relatively active live wildlife export trade across the country. Indeed, as early as the 1700s, it is reported that live chimpanzees were being exported from Sierra Leone,[51] while numerous European collectors travelled to the region during the 1800s to collect 'rare' and 'exotic' animals.[52] By the early 1900s, live animal exports started to become a major commercial enterprise in Sierra Leone. Sometime in the 1900s or 1910s, Reuben Castang, a famous dealer in animals and birds at London's Leadenhall Market, was sourcing chimpanzees from Sierra Leone. Three of these – Max, Moritz, and Akka – would go on to become the 'first Hollywood chimps', starring in numerous movies[53] and touring in theatres around the world. Their performances at the Tivoli Theatre in Melbourne, Australia, reportedly sold out for six weeks straight.[54]

Collecting animals also appeared to be popular with colonial residents in Sierra Leone: Charles Lane-Poole, the founder of the Forestry

Figure 5.2 Poster of Max and Moritz advertising their performances in Germany, ca. 1913.

Department, kept a large menagerie of animals at his residence in Freetown – even writing a short story about his pet Lucy the Chimp,[55] which he would recount to his children and grandchildren in later years.[56] In the late 1920s, Henry Trefflich, a famed Animal dealer from New York, set up an animal export outpost in Freetown, where he then proceeded, for the next twenty years, to export hundreds of chimpanzees among other animals. One of Trefflich's biggest customers was NASA, with many Sierra Leonean chimpanzees being used in the Agency's flight-testing programmes.[57] Thus, during the first half of the twentieth century, while there was little interest in enacting formal wildlife conservation programmes, there was certainly a great interest in the country's wildlife potential as a commercial commodity.

In the 1950s and 1960s, interest in Sierra Leone's wildlife had spread to zoos and documentary crews. In 1954, an expedition to Sierra Leone was organised by the British Broadcasting Corporation (BBC) and London Zoo, with the combined objectives of securing animals for the Zoo's collection and creating a documentary for audiences back in the United Kingdom. One of the expedition members was the then young BBC naturalist David Attenborough – it was his first time ever in Africa.[58] The final documentary made from the expedition, *Zoo Quest*, which Attenborough co-presented, was a widespread success and ultimately played a major role in launching Attenborough's career. In 1965, another expedition by a BBC film crew headed to Sierra Leone, this time to collect animals for Jersey Zoo. This expedition was led by the famed naturalist Gerald Durrell. Once again, the BBC documentary series and the eponymous book from the expedition, *Catch me a Colobus*, proved to be major successes.[59]

During this period of celebritisation of Sierra Leone's wildlife, the animal export industry started to develop more, driven by the demand from international animal collectors, zoos and, increasingly, medical laboratories.[60] Two men, in particular, became the main agents of Sierra Leone's animal export trade, Suleiman Mansaray and Franz Sitter. Mansaray, a Sierra Leonean national, appears to have taken over Trefflich's old animal export business in the 1950s, while Sitter, originally an Austrian national, arrived in Sierra Leone sometime during the 1950s. His history prior to his arrival has been the centre of much debate, with many people claiming that he was a former Nazi who fled to Sierra Leone after the fall of the Third Reich.[61] Mansaray specialised in birds, while Sitter specialised in reptiles, but both men dealt in primate exports, which proved to be the most lucrative trade of all.[62] Various sources estimate that between them they exported around 2,000 chimpanzees during the 1970s alone, most of them destined for medical testing laboratories in the United States, meaning that the vast majority of chimpanzees used in medical testing during this period would have originated from Sierra Leone.[63]

Wildlife Conservation Movement

After a long hiatus since Haywood's visit in the 1930s, attempts to regulate Sierra Leone's Wildlife population was reignited in the 1960s. Just after independence, in 1961, at the suggestion of the Colonial Office, the Sierra Leonean Government set up a 'Committee for the Preservation of Flora and Fauna' under the jurisdiction of the Ministry of Natural Resources.[64] In 1963, the FAO and IUCN sent a consultant to evaluate the current state of the country's wildlife.[65] The consultant, P.R. Hill, recommended that the government should enact a process of converting Forest Reserves into Game Reserves and set up a Wildlife Conservation Branch under the Forestry Department – essentially the same recommendations that Haywood had made thirty years earlier. Hill also identified the Tambakha Chiefdom in the north-east of the country as being an excellent site for Sierra Leone's first National Park. The Gola Forest was being logged by FIC at the time and was no longer seen to be an ideal site for the protection of wildlife.[66] Hill also noted the lively industry of primate exports. He did not, however, perceive the industry as being problematic for wildlife conservation; on the contrary, he described the efforts of the enterprise as being 'commendable'. His main concern instead was that revenue realised from the trade was not being effectively dispersed, and he therefore recommended that the government should enact a more rigorous taxation system for live animal exports.[67]

Hill's report was influential and evidently set the agenda for the Sierra Leonean Government's wildlife policy for the next decade and a half. It was also reflective of an important transitional moment in Sierra Leone, whereby international NGOs, rather than the British Colonial Government, started to become more involved in influencing Sierra Leone's forestry policy.[68] The Sierra Leonean Government immediately initiated the process of establishing a National Park in the north-east of Sierra Leone – to be called Outamba-Kilimi, while in 1967 a Wildlife Branch was established under the Forestry Department.[69] In 1972, a Wildlife Conservation Act was passed, and along with it four of the country's Forest Reserves were formally converted into 'Non-Hunting Forest Reserves'. While there were changes in legislation, overall there seemed to be little enthusiasm within the Forestry Department for wildlife initiatives. Indeed, during the 1960s and the 1970s, its two heads – Joshua Sawyerr and Musa Feika – tended to emphasise wildlife's commercial potential as a key factor.

> The principle underlying preservation is *not* that of a sentimental desire to protect wild animals from harm. It is based on a dispassionate appreciation of the facts. Wildlife affords a valuable aesthetic, educational and scientific to man's welfare and enjoyment. *Further, many species are of direct economic benefit.*[70]

> The government recognises that wildlife of Sierra Leone is a valuable natural resource of *economic importance* to the country and of territorial, international and cultural and scientific importance. It is the policy of Government to ensure the preservation and *rational utilization of wildlife* and its habitat for the best real benefit of the present and future [presumably human] inhabitants of Sierra Leone.[71]

> He [Sawyerr] said that local hunters were the real problem because they lived in remote places and were hard to control. 'We aren't bothered by the hunter from abroad, who does it for sport. He shoots one elephant, for which he will pay in hard currency and then go away. It's our own people who are the danger'.[72]
>
> Wild animals are an integral part of our ecosystem and it is the intention *to use them for the benefit of the people and the country*. They are a highly valuable food source and also yield hides, skins and other trophies.[73]

The above rhetoric of 'economic importance' appears to have been largely focused on governmental support for the live animal export trade, which, thanks to new taxes and fees introduced during the mid 1960s, was bringing in an important stream of revenue for the Government.[74] The Forestry Department claimed that the export of animals was on a quota basis, so that the number of animals leaving the country was controlled.[75] However, as one critic noted, these export quotas were not based on any knowledge of field populations or from capture results.[76] The commercial sports hunting of Sierra Leone also appears to have increased during the 1970s – the manager of the SILETI timber exploitation company in the Gola Forest reportedly made a name for himself as an accomplished hunter during his tenure, while in the late 1970s, a Lebanese pop singer reportedly travelled to Sierra Leone and shot dead 40–50 elephants.[77] Therefore, despite the passing of the Wildlife Conservation Act and tentative plans for a National Park in the north of Sierra Leone, the Forestry Department was more focused on regulating commercial wildlife ventures than developing a broader programme to ensure nationwide wildlife conservation.

The Primatologists

The late 1970s and the early 1980s represented an important juncture in Sierra Leone's wildlife conservation history, with the rise of a nascent wildlife conservation movement in-country, challenging the government's existing approaches to wildlife. Two expatriate primatologists – Dr Geza Teleki and Dr John Oates – who both separately visited Sierra Leone in 1979 for the first time, became important individuals in shaping Sierra Leone's future conservation policy and programmes. The two adopted distinctive approaches to promoting forest conservation: Teleki travelled to Sierra Leone as an activist, while Oates' initial priority was science; Teleki negotiated with the President to achieve his outcomes, while Oates had a greater focus on negotiating with local communities; Teleki focused his attention on wildlife conservation in the north-west savanna region of Sierra Leone, while Oates worked in the south-east in the tropical moist forest region; and, finally, Teleki worked with an existing NGO, while Oates was involved in setting up a new local environmental NGO. Both of them left their imprint on Sierra Leone's wildlife conservation movement.

Prior to both Teleki and Oates' arrival to Sierra Leone, a nascent wildlife conservation movement was being established by Sierra Leonean Daphne Tuboku-Metzger. Belonging to a prominent Krio (freed-slave descendent) family, Tuboku-Metzger had spent much of her early childhood being schooled in England. During this time (around the mid 1960s) she wrote a letter to the Sierra Leonean Government asking it to set up wildlife reserves.[78] In 1972, she completed a degree in education at the University of Durham and returned to Sierra Leone soon afterwards to take up teaching roles in a number of primary and secondary schools.[79] Her interest in wildlife issues, however, evidently never waned, and in 1976 she formally established Sierra Leone's first environmental NGO – the Sierra Leone Environment and Nature Conservation Association (SLENCA).[80] During its initial years, Tuboku-Metzger, along with another co-founder, personally funded the majority of the NGO's operations.[81]

The organisation quickly succeeded in gaining a profile in the local and international media and focused on curtailing the country's wildlife exports and setting up wildlife sanctuaries.[82] Notably, SLENCA was involved in establishing the Mamunta-Mayosa Nature Reserve in 1977 – the first wildlife reserve in Sierra Leone, although it never received any formal declaration under government laws.[83] SLENCA also had some early success in lobbying Sierra Leone's then President Siaka Stevens on wildlife exports. Stevens agreed to place a moratorium on chimpanzee exports in 1978 and accepted the position of being a patron of SLENCA. The declaration of the 'moratorium', however, proved to be largely symbolic, and within months of the export ban, eight chimpanzees shipped from Sierra Leone were confiscated at Schiphol Airport in Amsterdam due to a lack of import permits. The chimpanzees had been shipped by Franz Sitter, and the incident ultimately helped to focus international attention on Sierra Leone's trade in wildlife exports, indirectly resulting in the commissioning of Teleki's visit to Sierra Leone in 1979.[84]

Tuboku-Metzger remained a key figure in the Sierra Leonean wildlife conservation movement throughout the early 1980s, working closely with Teleki in particular as well as getting involved in a number of wildlife research projects.[85] By the late 1980s, however, SLENCA (now renamed 'Council for the Protection of Nature') and Tuboku-Metzger became less active in the wildlife conservation movement. It seems that Tuboku-Metzger focused more on her teaching career, heading a prominent independent school in Freetown. Her interest in environmental issues never faded though, and in 1999 she published a children's book (which was published in English and French) on endangered animals around the world.[86]

Geza Teleki: Chimpanzee Exports and the Outamba-Kilimi National Park

In 1979, Dr Teleki, then based at George Washington University, was asked by the United States Government to provide advice on the trade of chimpanzees being imported into medical laboratories. Teleki had a background in primate protection activities. He sat on the board of the International Primate Protection League (IPPL) and had previously worked on chimpanzee conservation programmes with Jane Goodall in East Africa. After some desk-based research, Teleki was surprised to find that almost all of the chimpanzees being used for medical testing in the United States were not only coming from 'one small country in West Africa' – Sierra Leone – but also from one dealer, Franz Sitter.[87]

He therefore proceeded to secure funds to finance a fact-finding mission and then travelled to Sierra Leone in late 1979, where he teamed up with Tuboku-Metzger and SLENCA to conduct a survey of the chimpanzee population in Sierra Leone and evaluate the impact of the primate export industry. The findings were designed to influence Sierra Leone's policy and feed into broader international campaigns for primate protection.[88]

Teleki's findings painted a dismal picture of wildlife conservation in Sierra Leone, and from them he deduced that there were only 2,000 chimpanzees left in Sierra Leone and that the operations of Sitter, along with those of Mansaray, had decimated the country's primate population. Teleki took the findings to President Siaka Stevens, and a process of negotiation to improve the conservation of the country's wildlife ensued. Stevens had been a supporter of Teleki's visit and the Forestry Department's wildlife conservation in general. The Sierra Leonean Government had helped to finance Teleki's in-country research, chimpanzee exports had been banned in 1978 (but poorly enforced), and Stevens had become the patron of SLENCA. Nevertheless, according to Teleki, Stevens' support for wildlife conservation was 'purely mercenary'; he was interested in wildlife conservation in so far as its ability to bring money into the country and improve the image of his government. Thus, while (discursively) Siaka Stevens promoted himself as being pro-wildlife conservation, in terms of actions, live animal exports continued and progress towards establishing Game Reserves and National Parks stagnated. Teleki recalled Stevens as being a person with great charisma: 'you knew you were talking to a crook, but at the same time it was an enjoyable conversation ... it was difficult to be upset or angry with him'.[89] The negotiations reached a discrete compromise; Siaka Stevens would enforce an overall ban on animal exports if Teleki was able to source overseas funding to make up for the loss in governmental revenue and finance the establishment of Sierra Leone's first National Park.[90]

The National Park was to be in the Tambakha Chiefdom in the north-east of Sierra Leone. It was the site that had been identified by Hill back in 1963,[91] but progress in converting it into a National Park had, for the previous decade and a half, been limited to a few pronouncements in the Government Gazette and the occasional survey.[92] Politically, it was also a safe site for a National Park as far as Siaka Stevens was concerned. The area had low soil fertility and therefore limited agriculture potential. It was also well outside the country's forestry and mining operations, and therefore would not compromise Government revenue – formal and informal.[93] The site of the proposed Park during the 1970s had become the focus of Sitter's animal hunting and operations. Missionaries had also reportedly been involved in regular elephant hunting expeditions in the area;[94] and therefore, to Teleki, the 'conservation of Outamba and Kilimi [was] a matter of life and death'.[95]

Teleki managed to source funding from the International Union for the Conservation of Nature (IUCN) and the World Wildlife Fund (WWF) for the National Park and continued to pressure Stevens into providing more political support. Teleki also acted as the director of the park project between 1981 and 1984, and during this period a considerable amount of research on wildlife was conducted in the proposed park's area.[96] Important for the National Park agenda was Sierra Leone's upcoming hosting of the Organisation of African Unity (OAU) conference (in 1981/2).[97] Prior to the OAU conference, the Stevens government had been criticised both internationally and regionally for its human rights abuses, its undemocratic reforms and overall poor governance. Stevens needed to construct some 'positive' aspect of his government to present during the conference. Teleki recognised this and unapologetically appealed to Stevens' hubris to gain further support for the National Park.[98] As his letters to the President reveal:

> Perhaps the suggestion that Your Excellency makes the announcement of an Outamba-Kilimi National Park at the upcoming OAU conference would not be out of order, as I am sure that such a statement would be applauded by many world leaders.[99]
>
> I therefore urge Your Excellency to extend a total ban to exports of *all* wildlife until an adequate number of Game Reserves and National Parks have been established and developed with financial aid and expert advice from overseas organizations … I respectfully submit that Your Excellency considers a formal statement to this effect being made at the upcoming OAU conference in Freetown, as I am sure such a step would bring praise from the leaders of other African nations where wildlife is strictly protected, and also from overseas governments and conservation organizations backing wildlife protection throughout Africa. The large amount of correspondence Your Excellency received concerning the chimpanzee exports issue is merely a small indication of the widespread international concern being shown for these matters, and I know the swift formation of one

> or two National Parks in Sierra Leone would be openly welcomed around the world.[100]

Teleki, it appears, had some success in influencing the President; Stevens approved all of the OKNP proposals, and in 1982 a ban on all wildlife exports was enacted.[101] The latter had some effect, and it appears that the wildlife exporter Mansaray went out of business around this time; Sitter's operations, however, persevered, albeit on a much more limited scale.[102] Teleki personally worked on establishing Outamba-Kilimi until 1984; however, he fell short of being able to have the area declared a National Park. The main issue delaying its formation was the financing of the resettlement and compensation of several hundred people living within the boundary of the Park's proposed boundaries.[103] Teleki left Sierra Leone that year, frustrated with local politics and the slow progress in achieving desired outcomes. He went on to advocate for global laws to stem the trade in live animals.[104]

After he left Sierra Leone, Teleki's initial work would go on to have a number of broader impacts, nationally and internationally, on wildlife conservation initiatives. His attempt at eliminating the export of chimpanzees was frustrated by the persistence of Franz Sitter's operations. In 1983, despite the animal export ban, fifty Sierra Leone chimpanzees were exported to a Japanese pharmaceutical company by Sitter. It appears that the exports were allowed in exchange for an aid shipment of rice from the Japanese Government.[105] When the Forestry Department's Chief Conservator of Forests, Musa Feika – who signed off on the export (and all previous animal exports for the past decade) – was challenged by the Convention on International Trade in Endangered Species of Wild Fauna and Flora (CITES) secretariat, he reportedly replied that 'to every rule their [sic] is an exception'.[106] Also in 1983, the Austrian multinational pharmaceutical company IMMUNO started to develop plans to set up a medical station in Sierra Leone that would conduct research on sixty to eighty chimpanzees. The plan, it appears, was a strategy to circumvent Austria's recent signing of CITES, which would have stymied chimpanzee imports for IMMUNO's research station in Vienna. Klaus Bieber, the honorary Austrian consul in Freetown and former business partner of Franz Sitter, was a central figure in negotiating an agreement with the Sierra Leone Government on behalf of IMMUNO, with Sitter being earmarked as the potential supplier of chimpanzees. The Sierra Leonean Government responded positively to the proposal, although it was evidently mindful of the international reaction to the enterprise.[107]

The International Primate Protection League (IPPL) found out about IMMUNO's proposal, and its director, Shirley McGreal, subsequently published a letter to the editor in the *Journal of Medical Primatology* criticising the proposal.[108] A large-scale campaign against IMMUNO was subsequently enacted by a variety of animal rights organisations.[109] IMMUNO reacted aggressively, suing McGreal and many others for libel. The case would finally be settled in 1989, against IMMUNO, and was seen as a famous case of upholding first amendment rights in the United States, where 'letters to the editor' were enshrined to be part of the right to free speech.[110] IMMUNO never established its research facility in Sierra Leone, finally giving up on the project in the early 1990s; Franz Sitter also left Sierra Leone around the same time, and thus Sierra Leone's role as a global supplier of medical testing chimpanzees also came to an end.[111] Teleki's campaign of eliminating the chimpanzee export trade out of Sierra Leone had finally succeeded, six years after he had left the country.

Progress on establishing Outamba-Kilimi as a National Park was also a slow affair. After Teleki left Sierra Leone, WWF was still involved in the overall management of the project and sourced a number of Peace Corps volunteers between the mid 1980s and 1992 to facilitate the park's management and development as well as secure its formal government gazettement.

Financing and the organisation of the resettlement for communities within the Park remained a perennial issue. The National Park's management and improvement in infrastructure were gradually upgraded throughout the 1980s; however, with the outbreak of Sierra Leone's civil war in 1991, funding sources dried up. Peace Corps volunteer Pam Seiser, who was working within the Park at the time, formally became a volunteer of the Forestry Department (as Peace Corps had left the country) to try and finalise the Park's establishment. However, after twelve months, due to a lack of funding and escalation of the conflict, Seiser left in 1993.[112] The National Park was finally (formally) declared in 1994, thanks to funding from a newly established coup-d'état Government in Freetown. However, it was a Pyrrhic victory, as rebels would later pass through the area destroying all of the Park's infrastructure.[113]

John Oates and Tiwai Island

John Oates first travelled to Sierra Leone in July 1979, almost a month before Teleki first arrived in the country to start his crusade against chimpanzee exports. Oates, at the time an anthropology faculty member at the City University of New York (CUNY), travelled to Sierra Leone to search for a site to research the Olive Colobus monkey. His efforts were rewarded when he came across Tiwai Island, on the Moa River, in the south-east of the country. It was a riverine island that not only contained the Olive Colobus monkey but also one of the highest densities of primates in the world.[114] Hunting practised by Liberian groups was common in the region as a result of the earlier monkey drives, and therefore Oates was concerned that their activities would soon be focused on Tiwai. He therefore saw it as essential that some form of Game Reserve should be established on the island to secure the wildlife's protection. Over the next decade, in conjunction with other scientists and Peace Corps Volunteers, he worked on establishing Tiwai Island as both a research centre and a tourism site.[115] Primate research thrived on the island,[116] and in 1987 it was formally declared as a Wildlife Sanctuary – the first area in Sierra Leone to be formally[117] protected for its wildlife.[118] The Sanctuary's declaration had resulted from a formal request made by the local chiefdom authorities of the island, and Tiwai Island was framed as a community-based wildlife conservation project, or at least one that had community involvement.[119]

In 1986, Oates, along with fellow primatologist Glynn Davies, was also involved in setting up the Conservation Society of Sierra Leone (CSSL). CSSL was, in part, set up in response to Davies and Oates' frustration with SLENCA, which operated on a membership by invitation basis. Tuboku-Metzger justified the policy on the basis of her suspicion of foresters and that their potential membership of SLENCA could undermine its objectives, as they were not fully committed to protecting wildlife. Oates and Davies argued that there was a need for an environmental NGO in Sierra Leone that could help to spark a more open forum for discussing the country's environmental future. Thus, with the help of a Peace Corps volunteer and some funding administered through WWF, CSSL was established.[120]

Around the same time, Glynn Davies was also commissioned to conduct a survey of the Gola Forest and comment on its potential for National Park status.[121] Since the beginning of the 1980s, there had been renewed calls to make parts of the Gola Forest a National Park or some form of wildlife reserve. The most vocal of these had come from the Sierra Leonean academic Norman Ayodele Cole, who, over the preceding couple of decades, had established himself as the foremost academic on Sierra Leone's environmental matters.[122] In 1980, Ayodele Cole argued that it was 'imperative that measures should be taken to declare suitable parts of the Gola Forest as biosphere reserves (or otherwise preserved areas) lacking any conflict with the economic development of the area'.[123] Progress by the Sierra Leonean Government in improving wildlife conservation in the area had nevertheless languished.[124] Davies' evaluation, however, helped to bring the Gola Forest into international prominence, and in 1990 the UK-based Royal Society for the Protection of Birds (RSPB),[125] in partnership with CSSL, decided to enact a programme to improve conservation in the Gola Forest. RSPB, initially at least, framed the programme as being beneficial to bird conservation in Europe:

> The link with West Africa is a simple one, as the majority of Western European breeding birds that spend the winter in Africa will pass through West Africa on their way farther south, or spend winter in the region. Conservation efforts in West Africa form a logical extension to the considerable effort which the RSPB is already making to improve bird conservation throughout western Europe.[126]

Overall, while progress was often faltering, between the late 1970s and early 1990s, a wildlife conservation movement and programme had emerged in Sierra Leone. During this period, the trade in live chimpanzee (and wild animal) exports had essentially been eradicated, plus Tiwai Island was established as a Wildlife Sanctuary, and the Gola Forest and Outamba-Kilimi were on track to become National Parks.

War and Peace in the Forests: 1990s to Present

For most of the 1990s, Sierra Leone was in a state of civil war, and this proved to be a major setback for wildlife conservation programmes in the country. The Gola Forest and Tiwai Island were some of the earliest areas to be affected, due to their close proximity to the Liberian border, where rebel militia first entered Sierra Leone.[127] John Oates, in his text *Myth and Reality in the Rain Forest*, recounts in detail the frustration during this period, as the research and tourist activities on Tiwai Island had to cease, and then there was the eventual ransacking of the Sanctuary's offices and infrastructure by rebels. The conflict soon spread across the country, and eventually the offices and infrastructure at Outamba-Kilimi National Park were also looted and destroyed. During the latter part of the 1990s, field trips by NGOs and government officials outside of Freetown became near impossible to conduct, and all field conservation activities came to a halt. Even Freetown was not safe from the ravages of the conflict. During the 1999 occupation of the city by the rebels, Prince Palmer, the Head of the Forestry Department, was killed. Palmer was identified by those who worked with him as being very intelligent and capable, as well as a strong supporter of wildlife conservation initiatives (establishing Tiwai Island as a Wildlife Sanctuary was his idea).[128] Some progress in chimpanzee conservation, however, was made during the conflict with the establishment of the Tacugama Chimpanzee Sanctuary on the outskirts of Freetown in 1995 by the expatriate Bala Amarasekaran.[129] While the live chimpanzee export trade had largely ceased during the early 1990s, there was still a widespread domestic trade in chimpanzees – notably in the form of pets for expatriate workers such as Peace Corps volunteers from the United States.[130]

Since the end of its civil war in 2001, there has been a great uptake in wildlife conservation activities. Wildlife conservation, to use the international development lexicon, has become a mainstream development activity, involving international NGOs and funding from major multilateral and bilateral aid donors. The Tiwai Island Wildlife Sanctuary was re-established under the management of the local environmental NGO Environmental Foundation for Africa (EFA), and since the mid 2000s research and tourism activities have resumed on the island.[131] The RSPB, in partnership with CSSL, and with major funding from the European Union, recommenced its Gola Forest programme, and in 2012 it succeeded in getting the forest re-gazetted as a National Park; while, a USAID/US Forest Service programme funded a large-scale conservation initiative for the Outamba Kilimi National Park. The Western Area Peninsula Forest Reserve (formerly known as the Colony Forest Reserve) was declared a National Park in June, 2013 (the country's third), largely thanks to a European Union funding project implemented by the German NGO Welthungerhilfe. The action of converting Forest Reserves into National Parks has been a major focus of conservation efforts in the country; the Loma Mountains Forest

Reserve, through a project being funded by the World Bank, was also in the process of being re-gazetted as a National Park,[132] while the Sierra Leonean Government has around a dozen more National Park gazettements planned for the near future.[133]

The actual historical change in wildlife population number in Sierra Leone, like with its forest cover, is hard to equate due to a lack of rigorous historical data. Elephant populations, however, which were estimated to be around 500 to 600 in the early 1900s,[134] have effectively been completely decimated, most likely due to hunting expeditions during the colonial and early independence eras. Leopard populations have also been decimated, most prominently during World War II, when, as mentioned, their skins became a popular commodity to sell to British servicemen based in Freetown. While the export of chimpanzees between 1950 and the mid 1980s was fairly rigorously documented (an estimated 5,000 were exported in total), historical numbers of chimpanzees in Sierra Leone before 1980 is unknown beyond speculative guesses. A 1932 report on wildlife in Sierra Leone, for example, just says that they are 'widely distributed but not plentiful'.[135] The Tacugama Chimpanzee Sanctuary conducted a Sierra Leone National Chimpanzee Census Project in 2010 and were surprised to find that 'the number of chimpanzees remaining in the wild exceeds 5,500', since a survey conducted Teleki and Baldwin in 1981 estimated that there were only between 1,500 and 2,500 chimpanzees in Sierra Leone. The 2010 survey was, however, coy in suggesting that there had been an actual increase of chimpanzees, instead suggesting flaws and limitations in the earlier survey being the problem. Notably, also, the survey found that 'more than half of [the chimpanzees surveyed] are to be found outside protected areas'.[136] Thus it appears that wildlife still manages to thrive outside of formally protected areas.

Wildlife Conservation Legacies

Although early on some laws, ideas and lobbying existed for wildlife conservation, it did not emerge as an institutionalised activity conducted by the Forestry Department until the postcolonial era. Furthermore, it offered a different philosophical and environmental perspective towards the forest. The Forest Department arguably wanted secure forests as places to source timber from (a resource-use ethics) and to a lesser extent for environmental services such as protecting watersheds, environmental concerns that are deeply anthropocentric in terms of their framing.[137] The wildlife conservation movement brought with it a new forest ontology: the forest as an important place for non-human species; the forest as a habitat and an ecosystem. Thus, the end-goals of forest and wildlife conservation differed considerably, although procedurally wildlife conservation in Sierra Leone built heavily on the discursive and material history of the colonial forestry department – that is, centralised control of forest spaces in Sierra Leone, building on a narrative of mistrust of local populations' competence in terms of managing and protecting forest spaces, and with a focus on 'reservationism' and converting 'Forest Reserves' into wildlife sanctuaries; this included changes to the Wildlife Conservation Act 1972, which converted Forest Reserves including the Loma Mountains, Tingi Gill, Kangari Hills and Western Area into non-hunting forest reserves and, more recently, the Gola Forest Reserves and the Western Area Peninsula Forest Reserve into National Parks. Indeed, with the establishment of National Parks, the wildlife movement achieved an early colonial forestry desire for much greater centralised control over the forest spaces. Thus, while the Wildlife Conservation movement might have different overall objectives in Sierra Leone, it builds strongly on colonial forestry legacies.

One of the most remarkable aspects of wildlife conservation is tracking the relationship between deforestation discourses and wildlife conservation objectives. As noted in previous chapters, the image of Sierra Leone as a heavily deforested country with a chronic degradation crisis epitomised the perspective of the Forestry Department, helping to frame its policies of forest reservation and afforestation since its inception in 1912. Interestingly, however, it appears that the discourse of deforestation was largely confined to the Forestry and Agricultural Departments during the colonial era, and it was not a widespread perception held by all colonial officials in Sierra Leone. Indeed, all of the district commissioners, when discussing wildlife in the 1930s, stated that the populations were stable and that there were no immediate threats; they discussed hunting with no mention of habitat destruction as a primary concern. Stanley – a northern commissioner and wildlife 'expert' – for example, noted that around Sierra Leone 'natural cover of every description is abundant',[138] an observation that contrasts with the perception of the Forestry and Agricultural Departments at the time, which described the country's vegetation as being in poor condition and under threat.[139] The deforestation crisis, it would seem, was therefore confined to the Departmental perspective and was not accepted across the entire colonial government.

Initial discussions about creating National Parks in the 1930s were therefore focused on their role in providing havens for animals from hunters – not as specific areas to arrest the spread of deforestation. It appears that the first time that the deforestation narrative is discussed in conjunction with wildlife conservation was is in 1963, with the visit of the international wildlife consultant P.R. Hill from the FAO/IUCN.[140] He stated that he had consulted previous Forestry Department reports as background reading to understand the country's vegetation. From this time onwards, reports and articles written by wildlife conservationists made references to Sierra Leone as once being heavily forested, making forest degradation a major concern for conservation initiatives.[141] Thus, it has only been since the 1970s that shifting cultivation and its associated deforestation has been discursively constructed as a major threat to Sierra Leone's wildlife.[142] Sierra Leone's early twentieth-century wildlife conservation history was rewritten, changing from Stanley's 1930s view of 'vegetation abundance' and 'healthy wildlife populations' to a period of 'dramatic alteration' of the environment.[143] The purpose of National Parks has also been reconstructed. They are not just seen as being wildlife safe havens from hunters but also as sites to arrest the country's deforestation, as a 1978 *International Primate Protection League Newsletter* alarmed: '[Sierra Leone] as yet has no National Parks and forests are being rapidly destroyed.'[144] Ultimately, in the 1970s and 1980s, the colonial historical degradation narrative was subsumed by the wildlife conservation movement in Sierra Leone.

The above suggests that Fairhead and Leach's West African 'deforestation narrative' needs to be understood with some nuance.[145] It was not a homogenised vision of colonial officials and later environmental actors, but rather a narrative that was utilised, ignored and shaped in different ways. This supports Swift's notion, as discussed in the previous chapter, that degradation narratives are important for making the government a stakeholder in environmental dilemmas.[146] The narrative was important for the Forestry Department to justify its forest reservation and afforestation programmes. However, the lack of interest in creating protected wildlife areas seems to have translated into a different reading of the landscape by early colonial officials like Stanley. Deforestation narratives only became important for Sierra Leonean wildlife conservation in conjunction with the rise of the National Park movement from the 1970s onwards.

A final aspect of Sierra Leone's wildlife conservation history, and to an extent it shares this with the broader colonial forest history, is the role of mainly expatriate individuals in shaping outcomes. Early on, elites from the Society for the Preservation of Fauna of the Empire tousled with the hunting enthusiast Governor Arnold Hodson to set early conservation policies, while later on, European primatologists laid foundations for wildlife sanctuaries and battled with animal exporters (including on alleged Nazi). Pioneer Sierra Leonean wildlife conservationist Daphne Tuboku-Metzger is a notable exception, although she recounts that her initial inspiration for wildlife conservation came from her early schooling years in England.[147] Wildlife conservation in Sierra Leone is thus, like the colonial forestry conservation ethic, arguably very much an imported ethic. This is not to suggest that local populations do no care for protecting wildlife, rather the mode and model of wildlife conservation in Sierra Leone closely aligns to a European model. In that sense, it shares some of the environmental dimensions of colonial forestry, an ontology of central government control over forest spaces, a favouring of (usually foreign) expert knowledge and implications for local forest user resource access. To paraphrase Martinez-Alier,[148] wildlife conservation is more of an environmentalism-of-the-elite, not an environmentalism-of-the-poor, and therefore wildlife conservation initiatives (e.g. establishment of National Parks; bans on hunting) have had implications for local populations' rights to access and use different forest-related resources. Although more recent iterations of wildlife conservation involving more participatory forms of local level participation have somewhat diluted the centralised approach, the establishment of large protected areas is still a core objective of the wildlife conservation movement.

Notes

1. The signatories of the convention, as listed, were 'The Governments of the Union of South Africa, Belgium, the United Kingdom of Great Britain and Northern Ireland, Egypt, Spain, France, Italy, Portugal and the Anglo-Egyptian Sudan.' Although a global convention, it was made clear that Africa was the focus, as the *1933 Convention Relative to the Preservation of Fauna and Flora in their Natural States* illustrates: 'Considering that the natural fauna and flora of certain parts of the world, and in particular of Africa are in danger, in present conditions, of extinction or permanent injury.'
2. A.H.W. Haywood, 'Sierra Leone: The Preservation of Wildlife', *Journal of the Society for the Preservation of the Wild Fauna of the Empire* 19 (1932), 21–32.

3. R.R. Granville, Replies to Questionnaire regarding Wild Life in the Northern Province of Sierra Leone, 1931: SLNA A/62/31.
4. A.H. Stocks, Questionnaire regarding Wildlife in Sierra Leone: Northern Province, 1931: SLNA – A/62/31.
5. W.B. Stanley, Memorandum, 1932: SLNA – A/62/31.
6. Haywood, 'Sierra Leone', 21–32.
7. D.K. Prendergast and W.M. Adams, 'Colonial Wildlife Conservation and the Origins of the Society for the Preservation of the Wild Fauna of the Empire (1903–1914)', *Oryx* 37(2) (2003), 251–260; R.P. Neumann, 'Dukes, Earls, and Ersatz Edens: Aristocratic Nature Preservationists in Colonial Africa', *Environment and Planning D* 14(1) (1996), 79–98.; R. Fitter and P. Scott, *The Penitent Butchers the Fauna Preservation Society, 1903–1978* (Collins, 1978).
8. CSP, Communication to HCS: 12 April, 1934: SLNA – A/33/28.
9. A. Hodson, *Trekking The Great Thirst: Travel and Sport in the Kalahari* (London: T F Unwin, 1912); A. Hodson, *Where Lion Reign: An Account of Lion Hunting and Exploration in South West Abyssinia* (Skeffington and Son Ltd, 1929).
10. A. Hodson, Communication to the Department of Zoology: 5 August, 1932: TNA – CO 267/636/14.
11. Hodson, Communication to the Department of Zoology – CO 267/636/14.
12. A. Hodson, Communication to the Natural History Museum, London: 12 February, 1932: TNA – CO 267/636/14.
13. Hodson, Communication to the Natural History Museum – CO 267/636/14.
14. *Field* – also known as *The Field, The Farm, The Garden, the Country Gentleman's Newspaper* (sporting paper) – was (and still is) a magazine dedicated to sports hunting; see M. Harris, *The Press in English Society from the Seventeenth to Nineteenth Centuries* (Associated University Presse, 1986).
15. G. Dollman, 'Dwarf Elephants' New Skull from the Gola Forest, *The Times*, June 27, 1935 p. 12.
16. Department of Zoology, Fauna of the Gola Forest, 1932: SLNA – A/15/32.
17. A. Fiddian, International communication within the Colonial Office: 4 March, 1932: TNA – CO 267/636/14; also see SH, Internationl Colonial Office Communication: 2 March, 1932: TNA – CO 267/636/14.
18. G. Dollman, Two Skulls of a Small Race of Elephant from the Cameroons and from Sierra Leone', *Proceedings of the Linnaean Society of London* (1933), 11–13.
19. Dollman, 'Dwarf Elephants', *The Times*, June 27, 1935, p. 12.
20. A. Hodson, Confidential Communication by the Governor: Protection of Fauna and Flora of Africa, 1933: SLNA – A/33/28.
21. Hodson, Confidential Communication by the Governor – A/33/28.
22. The *International Convention for the Protection of Fauna and Flora, 1936,* Article 7.

23. C.E. Cookson, Communication from Acting Governor to the Secretary of State for the Colonies (Sir Philip Cunliffe-Lister): August 18, 1933: SLNA – A/33/28.
24. HCS (JM), Communication to CSP: 18 April, 1934: SLNA – A/33/28.
25. The *International Convention for the Protection of Fauna and Flora, 1936.*
26. N.J. Jacobs, *Environment, Power, and Injustice: A South African History* (Cambridge University Press, 2003); N.J. Jacobs, 'The Intimate Politics of Ornithology in Colonial Africa', *Comparative Studies in Society and History* 48 (2006), 564–603; J. Carruthers, 'Tracking in Game Trails: Looking Afresh at the Politics of Environmental History in South Africa', *Environmental History* 11.4 (2006): 804–29; J. Carruthers, '"Police Boys" and Poachers: Africans, Wildlife Protection and National Parks, the Transvaal 1902 to 1950', *Koedoe* 36 (1993), 11–22; R.P. Neumann, 'The Postwar Conservation Boom in British Colonial Africa', *Environmental History* 7 (2002), 22–47; Neumann, 'Dukes, Earls, and Ersatz Edens', 79–98.
27. Carruthers, 'Tracking in Game Trails', 804–29; Neumann, 'The Postwar Conservation Boom', 22–47.
28. C.C. Gibson, *Politicians and Poachers: The Political Economy of Wildlife Policy in Africa* (Cambridge University Press, 1999).
29. K. Good, 'Settler Colonialism: Economic Development and Class Formation', *The Journal of Modern African Studies* 14 (1976), 597–620.
30. R.P. Neumann, *Imposing Wilderness: Struggles over Livelihood and Nature Preservation in Africa* (University of California, 1998).
31. P. Von Hellermann, *Things Fall Apart?: The Political Ecology of Forest Governance in Southern Nigeria* (Berghahn Books, 2013).
32. Neumann, *Imposing Wilderness.*
33. T.S. Jones, 'Sierra Leone', in P. Grubb, T.S. Jones, A.G. Davies, E. Edberg, E.D. Starin and J.E. Hill (eds), *Mammals of Ghana, Sierra Leone and the Gambia* (Trendrine Press, 1998), 27–45.
34. During this period, Sierra Leone was divided into six districts for administrative purposes.
35. A notable exception was the hunting expeditions conducted in Northern Sierra Leone by Lieutenant Wills in 1910; see G.E.J. Wills, 'After Elephants in the Sierra Leone Protectorate and French Guinea', *Baily's Magazine* 619 (1911), 184–91.
36. W.B. Stanley, 'Wild Life in Sierra Leone', *The Field* (1933), 1044.
37. T.F. Edwardson, Sierra Leone Diaries – July 4, 1935: RHA – Mss afr r 210(1); C.E. Lane-Poole, Sierra Leone Diaries – 25 and 28 March; 10, 15 and 28 April; 21 May, 1913: NLA – MS 3799/5; C.E. Lane-Poole, Sierra Leone Diaries – 15 January, 1914: NLA – MS 3799/5; C.E. Lane-Poole, Sierra Leone Diaries – 19 November, 1914: NLA – MS3799/6; *The Sierra Leone Royal Gazette*, Government Notices, May 2, 1925: 383.
38. W.B. Stanley, 'Carnivorous Apes in Sierra Leone', *Sierra Leone Studies* 2 (1919), 3–19; Stanley, Memorandum, 1932: SLNA – A/62/31; W.B. Stanley,

'Game Preservation in Sierra Leone', *Sierra Leone Studies* 11 (1933), 2–15; Stanley, 'Wild Life in Sierra Leone', 1044; W.B. Stanley and E. Hodgson, *Elephant Hunting in West Africa* (Geoffrey Bles, 1929).

39. A.H. Unwin, *Report on the Forest and Forestry Problems in Sierra Leone* (Waterlow and Sons Limited, 1909).
40. L. Palfreman, Communication to the Colonial Secretary: 17 August, 1918: SLNA A/69/30.
41. Colonial Secretary, Communication to the Conservator of Forests: 27 September, 1918: SLNA A/69/30.
42. District Commissioner Kabala, Communication to the Colonial Secretary: 17 August, 1918: SLNA A/69/30.
43. Colonial Secretary, Communication to District Commisioner Kabala: 18 September, 1918: SLNA A/69/30.
44. Stanley, Memorandum, 1932: SLNA – A/62/31; Stanley, 'Game Preservation', 2–15; Stanley, 'Wild Life in Sierra Leone', 1044.
45. Stanley, 'Game Preservation', 15.
46. *Forestry Ordinance Amendment, 1924.*
47. *Daily Mail*, 'Thousands of Monkeys Die in Four Days', 7 November, 1955, 7; Granville, Replies to Questionnaire, 1931: SLNA A/62/31.
48. FAO Release, 'Leopards Out, Monkeys In', *The Ecologist* 1 (1978), 32; R.W. Hayman, 'Mammals of Sierra Leone', *Zoo Life* 10 (1955), 2–6; P. Hill, *FAO/IUCN African Special Project: Interim Report on Sierra Leone* (IUCN, 1963); T.S. Jones, *Notes on the Monkeys of Sierra Leone* (Department of Agriculture, 1950); T.S. Jones, 'The Status of Large Mammals and their Conservation in Sierra Leone', paper presented at the Wildlife Conservation in Africa, Kew, 1993; T.S. Jones, 'Appendix II – The Sierra Leone Monkey Drives', in P. Grubb, T.S. Jones, A.G. Davies, E. Edberg, E.D. Starin and J.E. Hill (eds), *Mammals of Ghana, Sierra Leone and the Gambia* (The Trendine Press, 1998), 214–19. Jones, 'Sierra Leone', 27–45; R.H.G. Lowes, 'Destruction in Sierra Leone', *Oryx* 10(5) (1970), 309–10.
49. Jones, *Notes on the Monkeys of Sierra Leone*; Jones, 'Appendix II', 214–19.
50. A.G. Davies, *The Gola Forest Reserves, Sierra Leone: Wildlife Conservation and Forest Management* (IUCN, 1987); Jones, 'The Status of Large Mammals'; Jones, 'Appendix II', 214–19; Jones, 'Sierra Leone', 27–45; J.F. Oates, *Myth and Reality in the Rain Forest: How Conservation Strategies are Failing in West Africa* (University of California Press, 1999); P.T. Robinson, 'Wildlife Trends in Liberia and Sierra Leone', *Oryx* 11(2–3) (1971), 117–121; G. Teleki, *Hunting and Trapping Wildlife in Sierra Leone: Aspects of Exploitation and Exportation* (Special Report to MAF, 1980).
51. Jones, 'Sierra Leone', 27–45; G. Teleki and L. Baldwin, 'Sierra Leone's Wildlife Legacy', *Zoonooz* 54(10) (1981), 21–27.
52. Jones, 'Sierra Leone', 27–45.
53. For example, T. Bently (Writer) in J. Haimann (Producer), *No Monkey Business* (General Film Distributors, 1935).

54. *NZ Truth*, 'Animal Comedy', 2 January, 1928; *Sydney Morning Herald*, 'Famous Monkeys – Arrival From America', 15 July, 1927; R.W. Thompson, *Wild Animal Man: Being the Story of the Life of Reuben Castang* (W. Morrow & Company, 1934).
55. C.E. Lane-Poole, Lucy the Chimpanzee. Unpublished Manuscript, 1922: NLA – 3799/5.
56. J. Dargavel, *The Zealous Conservator: The Life of Charles Lane Poole* (University of Western Australia, 2008).
57. H. Trefflich and E. Anthony, *Jungle For Sale* (New York: Hawthorn Books, 1967); P.G. Munro, 'Geza Teleki and the Emergence of Sierra Leone's Wildlife Conservation Movement' *Primate Conservation* 29 (2015), 115–22.
58. D. Attenborough, 'Expedition to Sierra Leone', *Zoo Life* 10 (1955), 11–20.
59. G. Durrell (Writer) (1966) in C. Parsons (Producer), *Catch me a Colobus* (London: BBC – Natural History Unit, 1966); Durrell, *Catch Me A Colobus*.
60. *Daily Mail*, 'Sierra Leone Chimps for New York', 11 May, 1955, 6–7; Teleki, *Hunting and Trapping Wildlife in Sierra Leone*.
61. Geza Teleki, personal communication, 2012; John Oates, personal communication, 2011; D. Peterson and J. Goodall, *Visions of Caliban: On Chimpanzees and People* (Houghton Mifflin Company, 1993).
62. *International Zoo Yearbook*, International List of Animal Dealers 2 (1961), 240–48; Teleki and Baldwin, 'Sierra Leone's Wildlife Legacy', 21–27; Munro, 'Geza Teleki', 115–22.
63. *International Primate Protection League Newsletter*, 'Chimpanzee Traffickers Denounced in Sierra Leone Press' 3 (1978), 5; Peterson and Goodall, *Visions of Caliban*; J.S. Sawyerr, *Report on the Wildlife of Sierra Leone* (Forestry Department, 1963); G. Teleki, *Outamba-Kilimi National Park: A Provisional Plan for Management and Development* (Prepared for the government of Sierra Leone and the IUCN and WWF,1986).
64. Teleki, *Outamba-Kilimi National Park*.
65. Hill, *FAO/IUCN African Special Project: Interim Report on Sierra Leone*; P. Hill, 'Sierra Leone', in T. Riney and P. Hill (eds), *Conservation and Management of African Wildlife* (Rome: FAO, 1967), 72–97.
66. Lowes, 'Destruction in Sierra Leone', 309–10; A.F. Wilkinson, 'Areas to Preserve in Sierra Leone', *Oryx* 12(5) (1974), 596–98.
67. Hill, *FAO/IUCN African Special Project: Interim Report on Sierra Leone*.
68. W. Kontch, 'Forest Resource Management in Sierra Leone: A Critique of Policy Formation and Implementation', PhD thesis (University of Leeds, 1997).
69. Lowes, 'Destruction in Sierra Leone', 309–10. Sawyerr, *Report on the Wildlife of Sierra Leone*.
70. Sawyerr, *Report on the Wildlife of Sierra Leone*, 17.
71. J.S. Sawyerr, *Ten-Year Plan of Forestry Development for Sierra Leone: 1964/5 – 1973/1974* (Ministry of Natural Resources, 1964), ii.
72. FAO Release, 'Leopards Out, Monkeys In', 32.

73. M.B. Feika, *Report on the Forest Administration of Sierra Leone 1963/1964 to 1973/74* (Government Printer, 1975), 17.
74. Sawyerr, *Ten-Year Plan of Forestry Development*; J.S. Sawyerr, *Report on the Forest Administration of Sierra Leone for the year April, 1963 to March, 1964* (Government Printer, 1965).
75. Sawyerr, *Report on the Wildlife of Sierra Leone.*
76. Robinson, 'Wildlife Trends in Liberia and Sierra Leone', 117–21.
77. Peterson and Goodall, *Visions of Caliban*; D. Tuboku-Metzger, 'Nature Conservation in Sierra Leone', *African Wildlife News* 14(3) (1979), 12–16.
78. D. Tuboku-Metzger, *In Danger! Endangered Species of the World* (Heinemann Educational Books, 1999).
79. N. Polunin and M.L. Curme, *World Who is Who and Does What in Environment & Conservation* (Earthscan, 1997).
80. SLENCA has also been described as being one of the first indigenous wildlife conservation NGOs in Africa; see Polunin and Curme, *World Who is Who.*
81. *Oryx*, 'Conservationists in Sierra Leone', 14 (1977), 18.
82. *International Primate Protection League Newsletter*, 'Chimpanzee Traffickers Denounced', 5; Tuboku-Metzger, 'Nature Conservation in Sierra Leone', 12–16; D. Tuboku-Metzger, 'Robert Lowes, 1914–1990', *Oryx* 25(2) (1991), 118; WCMC, *Protected Areas of the World: A Review of National Systems. Volume 3: Afrotropical* (IUCN, 1991).
83. Polunin and Curme, *World Who is Who*; Tuboku-Metzger, 'Robert Lowes', 118.
84. *International Primate Protection League Newsletter*, 'Ten Chimpanzees Seized at Amsterdam Airport', 6(1) (1979), 6; Teleki, *Outamba-Kilimi National Park.*
85. E.g. R.R. Reeves, D. Tuboku-Metzger and R.A. Kapindi, 'Distribution and Exploitation of Manatees in Sierra Leone', *Oryx* 22 (1988), 75–84; Polunin and Curme, *World Who is Who.*
86. Tuboku-Metzger, *In Danger!*
87. Geza Teleki, personal communication, 2012.
88. *International Primate Protection League Newsletter*, 'News from Sierra Leone', 7(2) (1980), 14; *International Primate Protection League Newsletter*, 'Sierra Leone Chimpazees Threatened', 10(2) (1983), 6–7.
89. Geza Teleki, personal communication, 2012.
90. Geza Teleki, personal communication, 2012.
91. As early as 1931, the Colonial District Commissioner of the Northern Province had identified the Tambakha Chiefdom as a potentially good site for a Game Reserve or National Park; see Granville, Replies to Questionnaire, 1931: SLNA A/62/31.
92. H. McGriffin, 'History of Primate Conservation in Sierra Leone', *International Primate Protection League Newsletter* 12(3) (1985), 3–5; Teleki, *Outamba-Kilimi National Park*; G. Teleki and I. Bangura, 'Outamba-Kilimi National Park: Cornerstone for Conservation', *Zoonooz,* 54(10) (1981),

281–83; J. Waugh, *Decriptive Information for Outamba-Kilimi National Park* (Republic of Sierra Leone, 1986).

93. Paul Richards, personal communication, 2012; Geza Teleki, personal communication, 2012.
94. H. McGriffin, 'A New National Park in Sierra Leone', *International Primate Protection League Newsletter* 12(3) (1985), 2–3; G. Teleki, 'Report from Sierra Leone', *Traffic (International) Bulletin* 11 (1980), 78; Teleki, *Outamba-Kilimi National Park.*
95. *International Primate Protection League Newsletter*, 'News from Sierra Leone', 14.
96. G.T. Zug, 'Amphibians and Reptiles of the Outamba-Kilimi Region, Sierra Leone', *The Journal of the Herpetological Association of Africa* 33 (1987), 1–4; D.P. Harding and R.S.O. Harding, 'A Preliminary Checklist of Birds in the Kilimi Area of Northwest Sierra Leone', *Malimbus* 4 (1982), 64–68; R.S.O. Harding, 'Primates in the Kilimi Area, North-Western Sierra Leone', *Folia Primatologica* 42(2) (1984), 96–114; R. Happel, 'Seed-eating by West African Cercopithecines, with Reference to the Possible Evolution of Bilophodont Molars', *American Journal of Physical Anthropology* 75(3) (1988), 303–27; G. Teleki, A.G. Davies and J .Oates, 'Sierra Leone', in R. East (ed.), *Antelopes: Global Surveys and Regional Action Plans. Part 3: West and Central Africa* (IUCN, 1990), 40–46.
97. J.D. Kandeh, 'Ransoming the State: Elite Origins of Subaltern Terror in Sierra Leone', *Review of African Political Economy* 26(81) (1999), 349–366.
98. Geza Teleki, personal communication, 2012.
99. G. Teleki, Letter to President Siaka Stevens: 24 May, 1980: Mimeo.
100. Teleki, Letter to President Siaka Stevens.
101. H. McGriffin, 'History of Primate Conservation in Sierra Leone', *International Primate Protection League Newsletter* 12(3) (1985), 3–5.
102. Teleki, *Outamba-Kilimi National Park*; Geza Teleki, personal communication, 2012.
103. Harding and Harding, 'A Preliminary Checklist of Birds', 64–68; OKNP Staff, personal communication, 2012.
104. Geza Teleki, personal communication, 2012.
105. *Animals International*, 'Chimps in Danger – Sierra Leone', 11(19) (1986); A. Kabasawa, 'The Chimpanzees of West Africa. From "Man-Like Beast" to "Our Endangered Cousin"', in T. Matsuzawa, T. Humle and Y. Sugiyama (eds), *The Chimpanzees of Bossou and Nimba Primatology Monographs* (Springer, 2011), 45–57; *Traffic Bulletin*, 'Sierra Leone's Chimps Endangered by Commerical Exploitation', 5(3–4) (1983), 48.
106. Peterson and Goodall, *Visions of Caliban*, 110.
107. N. Heneson, 'Loophole May Allow Trade in African Chimps', *New Scientist*, 20 October, 1983, 165; *Traffic Bulletin*, 'Sierra Leone's Chimps', 48.
108. S. McGreal, 'Letter to the Editor: A Project with Potential to Spread Non-A, Non-B Hepatitis in West Africa', *Journal of Medical Primatology* 12 (1983),

280–82; also see J. Moor-Jankowski, 'Editorial Note: A Project with Potential to Spread Non-A, Non-B Hepatitis in West Africa', *Journal of Medical Primatology* 12 (1983), 280.

109. *Animals International*, 'Chimps in Danger'; *International Primate Protection League Newsletter*, 'Sierra Leone Chimpazees Threatened', 6–7; *Traffic Bulletin*, 'Sierra Leone's Chimps', 48.
110. IMMUNO AG v. Moor-Jankowski, No. 15 A.D.2d 114 (Supreme Court of the State of New York, 1989); *International Primate Protection League Newsletter*, 'Immuno Lawsuit Thrown Out – Twice More!' 17(1) (1990), 17–19; *International Primate Protection League Newsletter*, Immuno Sues – 1984', 20(3) (1993), 17; Peterson and Goodall, *Visions of Caliban*; *Scientific American*, 'PROFILE: Jan Moor-Jankowski: Whistle-Blower's Wars', (1997), 32–33.
111. A. Kabasawa, 'Current State of the Chimpanzee Pet Trade in Sierra Leone', *African Study Monographs* 30(1) (2009), 37–54; Peterson and Goodall, *Visions of Caliban*.
112. Pam Seiser, personal communication, 2012.
113. C. Squire, *Sierra Leone's Biodiversity and the Civil War: A Case Study Prepared for the Biodiversity Support Program* (World Wildlife Fund, 2001).
114. J.F. Oates, G.H.Whitesides, A.G. Davies, P.G. Waterman, S.M. Green, G.L. Dasilva and S. Mole, 'Determinants of Variation in Tropical Forest Primate Biomass: New Evidence from West Africa', *Ecology* 71(1) (1990), 328–43; G.H.Whitesides, J.F. Oates, S.M. Green and R.P. Kluberdanz, 'Estimating Primate Densities from Transects in a West African Rain Forest: A Comparison of Techniques', *The Journal of Animal Ecology* 57(2) (1988), 345–67.
115. Oates recounts the process of setting up Tiwai Island in detail in his text *Myth and Reality in the Rain Forest*; also see G. Dasilva, 'An Introduction to the Tiwai Conservation Project', *Primate Eye* 35 (1988), 24–25.
116. For example, see G. Dasilva, 'Diet of Colobus Polykomos on Tiwai Island: Selection of Food in Relation to its Seasonal Abundance and Nutritional Quality', *International Journal of Primatology* 15(5) (1994), 655–80; A.G. Davies, J.F. Oates and G.L. Dasilva, 'Patterns of Frugivory in Three West African Colobine Monkeys', *International Journal of Primatology* 20(3) (1999), 327–57; C. Fimbel, 'Cross-species Handling of Colobine Infants', *Primates* 33(4) (1992), 545–49; C. Fimbel, 'Ecological Correlates of Species Success in Modified Habitats May Be Disturbance- and Site-Specific: The Primates of Tiwai Island', *Conservation Biology* 8(1) (1994), 106–13; C. Fimbel, 'The Relative Use of Abandoned Farm Clearings and Old Forest Habitats by Primates and a Forest Antelope at Tiwai, Sierra Leone, West Africa', *Biological Conservation* 70(3) (1994), 277–86; J.F. Oates, 'The Diet of the Olive Colobus Monkey, Procolobus Verus, in Sierra Leone', *International Journal of Primatology* 9(5) (1988), 457–78; J.F. Oates and G.H. Whitesides, 'Association between Olive Colobus (Procolobus Verus),

Diana Guenons (Cercopithecus Diana), and Other Forest Monkeys in Sierra Leone', *American Journal of Primatology* 21(2) (1990), 129–46; Oates et al. 'Determinants of Variation in Tropical Forest', 328–43; G.H. Whitesides, 'Nut Cracking by Wild Chimpanzees in Sierra Leone, West Africa', *Primates* 26(1) (1985), 91–94.; G.H. Whitesides, 'Interspecific Associations of Diana Monkeys, Cercopithecus Diana, in Sierra Leone, West Africa: Biological Significance or Chance?' *Animal Behaviour* 37 (1989), 760–76; Whitesides et al., 'Estimating Primate Densities', 345–67.

117. Mamunta-Mayosa Nature Reserve had been 'informally' declared 10 years earlier.
118. E.K. Alieu, 'People's Participation in the Development and Management of Tiwai – Sierra Leone's First Game Sanctuary', *Wildlife and Nature* 11(3) (1995), 11–21.
119. J. F. Oates, 'Sierra Leone's New Sanctuary', *Oryx* 22(1) (1988), 50–51; Oates, *Myth and Reality in the Rain Forest.*
120. John Oates, personal communication, 2011; Glynn Davies, personal communication, 2011; Oates, *Myth and Reality in the Rain Forest.*
121. Davies, *The Gola Forest Reserves*; A.G. Davies and P. Richards, *Rain Forests in Mende Life: Resources and Subsistence Strategies in Rural Communities around the Gola North Forest Reserve (Sierra Leone)* (ESCOR, UK Overseas Development Administration, 1991).
122. N.H.A. Cole, 'Ecology of the Montane Community at the Tingi Hills in Sierra Leone', *Bulletin de L'IFAN* 29(3) (1967), 904–23; N.H.A. Cole, 'Ecology of a Moist Semi-deciduous Forest on Kogia Hill in Sierra Leone', *Bulletin de L'IFA* 30(1) (1968), 100–13; N.H.A. Cole, 'Review of Classification of Vegetation in Sierra Leone', *West African Science Association Journal* 13(1) (1968), 81–92; N.H.A. Cole, *The Vegetation of Sierra Leone* (Njala University Press, 1968); N.H.A. Cole, 'Climate, Life Forms and Species Distribution on the Loma Montane Grassland, Sierra Leone', *Botanical Journal of the Linnean Society* 69(3) (1974), 197–210; N.H.A. Cole, 'Toward Solving Problems of Environmental Stress in Developing Countries', *Environmental Management* 3(6) (1979), 479–82; N.H.A. Cole, 'The Gola Forest in Sierra Leone: A Remnant of Tropical Rain Forest in Need of Conservation', *Environmental Conservation* 7(1) (1980), 33–40; N.H.A. Cole, 'Floristic Associations in the Gola Rain Forests: A Proposed Biosphere Reserve', *Journal of Pure and Applied Sciences* 2(1) (1995), 35–45; N.H.A. Cole, 'The Role of Biodiversity Conservation in Implementing the Tropical Forestry Action Plan in Sierra Leone', *Journal of Pure and Applied Sciences* 4 (1995), 31–40.
123. Cole, 'The Gola Forest in Sierra Leone', 40.
124. Cole, 'Floristic Associations in the Gola Rain Forests', 35–45; Cole, 'The Role of Biodiversity Conservation', 31–40.
125. The Royal Society for the Protection of Birds (RSPB) is one of the oldest conservation organisations in the world. It was established in 1891 and was largely led by women opposed to the plumage trade that was causing

the extinction of certain birds; see S. Jones, 'A Political Ecology of Wildlife Conservation in Africa', *Review of African Political Economy* 33(109) (2006), 483–95.

126. N. Coulthard, 'The RSPB has a New Project in Africa', *Birds Magazine* Winter (1990), 18.
127. M. Leach, *Rainforest Relations: Gender and Resource Use among the Mende of Gola, Sierra Leone* (Edinburgh: Edinburgh University Press, 1994); P. Richards, 'The Sierra Leone-Liberia Boundary Wilderness: Rain Forests, Diamonds and War', in A. Asiwaju and P. Nugent (eds), *African Boundaries: Constraints, Conduits and Opportunities* (Pinter, 1996), 205–21.
128. John Oates, personal communication, 2011; Glynn Davies, personal communication, 2011.
129. Kabasawa, 'Current State of the Chimpanzee Pet Trade', 37–54; A. Kabasawa, R.M. Garriga and B. Amarasekaran, 'Human Fatality by Escaped Pan Troglodytes in Sierra Leone', *International Journal of Primatology* 29 (2008), 1671–685.
130. Kabasawa, 'Current State of the Chimpanzee Pet Trade', 37–54.
131. P.G. Munro, 'Protected Area Management and Community Development in Tiwai', *Africa News*, 27 July, 2007; A.J. Sundufu, M.S. James, I.K. Foday and T.F. Kamara, 'Influence of Community Perceptions towards Conservation and Eco-tourism Benefits at Tiwai Island Wildlife Sanctuary, Sierra Leone', *American Journal of Tourism Management* 1(2) (2012), 45–52.
132. D.T. Jumpah, *Bumbuna Hydroelectric Environmental and Social Management and Biodiversity Conservation Project* (Government of Sierra Leone, 2012).
133. John Oates, personal communication, 2011.
134. Haywood, 'Sierra Leone', 21–32.
135. Haywood, 'Sierra Leone', 21–32.
136. T.M. Brncic, B. Amarasekaran, and A. McKenna, *Sierra Leone National Chimpanzee Census September 2010* (Tacugama Chimpanzee Sanctuary, 2010).
137. For example, see R. Eckersley, *Environmentalism and Political Theory: Toward an Ecocentric Approach* (SUNY Press, 1992); A. Dobson, '"Environmental Sustainabilities": An Analysis and a Typology', *Environmental Politics* 5 (1996), 401–428.
138. Stanley, 'Game Preservation', 15.
139. See E. MacDonald, *Sierra Leone Annual Report of the Forest Department for the Year 1931* (Government Printer, 1932); E. MacDonald, *Sierra Leone, Annual Report of the Forest Department for the Year 1932* (Government Printer, 1933); E. MacDonald, *Sierra Leone Annual Report of the Forest Department for the Year 1934* (Government Printer, 1935); F.J. Martin, *A Preliminary Survey of Vegetation of Sierra Leone* (Department of Agriculture, 1937).
140. Hill, *FAO/IUCN African Special Project: Interim Report on Sierra Leone*; Hill, 'Sierra Leone', 72–97.

141. Teleki, *Outamba-Kilimi National Park*; Teleki, *Hunting and Trapping Wildlife in Sierra Leone*; J.F. Oates, *Action Plan for African Primate Conservation: 1986–1990: IUCN/SSC Primate Specialist Group* (UNEP and IUCN, 1985); Oates, *Myth and Reality in the Rain Forest*; Tuboku-Metzger, 'Nature Conservation in Sierra Leone', 12–16; D. Tuboku-Metzger, 'Forest Exploitation in Sierra Leone: A Tale of Devastation', *The Ecologist* 13(6) (1983), 239–41.
142. Harding and Harding, 'A Preliminary Checklist of Birds', 64–68; Harding, 'Primates in the Kilimi Area, North-Western Sierra Leone', 96–114.
143. Teleki, *Hunting and Trapping Wildlife in Sierra Leone*, 1.
144. *International Primate Protection League Newsletter*, 'Sierra Leone Wildlife Laws', 5 (1978), 15.
145. J. Fairhead and M. Leach, 'Reconsidering the Extent of Deforestation in Twentieth-Century West Africa', *Unasylva* 192 (1998), 38–46.
146. J. Swift, 'Desertification: Narratives, Winners & Losers', in M. Leach and R. Mearns (eds), *The Lie of the Land: Challenging Received Wisdom on the African Environment* (Oxford: James Curry, 1996), 73–90.
147. Tuboku-Metzger, *In Danger!*
148. J. Martinez-Alier, *The Environmentalism of the Poor: A Study of Ecological Conflicts and Valuation* (Edward Elgar Publishing, 2003).

Epilogue

In March 2015, Vice President Samuel Sam-Sumana was controversially expelled from his political party and sacked from his position as Vice President, which, among other things, ended his annual hikes up the Western Area Peninsula Mountains in support of National Tree Planting Day. Somewhat ironically, his political demise was in part linked to allegations surrounding the cutting down of trees.

Initially selected in 2007 by the then presidential candidate Ernest Bai Koroma to act as a running mate, Sam-Sumana's main campaign role was to shore up support in his home district, Kono, a critical 'swing' district in Sierra Leonean national elections.[1] His subsequent tenure as Vice President, however, was marred with numerous controversies,[2] and his eventual falling out with his political party (and his loss of his Vice Presidency) in 2015 was climaxed with Sam-Sumana seeking political asylum in the United States embassy in Freetown. He claimed he was being persecuted due to his lack of political party membership.[3] This falling out from his political party, however, had deep roots, and even prior to the 2012 national elections in Sierra Leone many within the party reportedly had wanted Samuel Sam-Sumana dropped from the Vice Presidency post.[4]

The 2012 internal party rumblings around Sam-Sumana's Vice Presidency had emerged in the aftermath of two timber export controversies. Most notably, in 2011, an Al Jazeera investigative documentary titled *Timber!* presented footage of two of the Vice President's aides requesting bribes (from undercover journalists) to secure the Vice President's support for a timber export business. [5] This was at a time when timber exports were banned in Sierra Leone. The meeting took place in the Vice President's office, and a US$2,000 bribe was accepted. The two men were later charged (although the Vice President was not), and the whole affair became popularly known by the Nixonesque moniker 'Timbergate'. Also, in mid 2012, it was revealed via a WikiLeaks US embassy cable[6] that between 2006 and 2008 the Vice President had been involved in setting up a company with US-based businessmen to commence logging concessions in the north-east of Sierra Leone. The venture collapsed, and the US-based investors subsequently claimed that the commercial investments wired to Sam-Sumana had been diverted into campaign funds for the Vice President's 2007 election bid. The Vice President was even named as a third-party defendant in a civil law suit in the United States in relation to money borrowed for the venture. One of the Vice President's 'Timbergate' aides was also allegedly caught up in this scandal.[7] To an extent, the Vice President managed to survive these scandals: the Sierra Leonean anti-corruption commission found no direct evidence that the Vice President was aware of the illegal timber-related bribe in his office,[8] while the United States lawsuit was eventually dropped. Nevertheless, both undoubtedly resulted in political damage for the Vice President, undermining his support in his political party and likely his standing among the voting populace.[9]

Beyond the travails of the electoral politics of these incidents, the WikiLeaks-related scandal in particular provides some additional background and insights through which to view the Vice President's annual hikes up the Western Area Peninsula Mountains, which he conducted ostensibly in support of tree planting and conservation. The WikiLeaks cable,[10] along with investigative reports,[11] revealed that Sam-Sumana held a 10 per cent stake in the company Taakor Tropical Hardwoods, which was set up with Missouri-based businessman Ed Sandridge. The company was established in 2006 in both the United States and Sierra Leone. It was given a major logging concession in the Tama-Tonkolili Forest Reserve in 2007 and soon after it commenced harvesting operations. However, Taakor Tropical Hardwoods was unable to export the logs due to a temporary government ban on timber exports that was implemented in 2008. The entire stock of harvested logs that was 'ready for export was then either stolen or it rotted away'.[12] What was revealing from this episode, in addition to the court-related scandals, was that Vice President Sam-Sumana had been secretly (i.e. he did not disclose the information to the public) involved in the timber export business. Indeed, it is alleged that he even bribed media sources to try and bury the story prior to the WikiLeaks revelations.[13] Overall, it is perhaps an unsurprising revelation given Sam-Sumana's pre-political career in the diamond export business, but it sheds new light on his environmental treks up the Western Area Peninsula Mountains. While he was discussing the need for Forest Reserves, the importance of wildlife conservation and symbolically planting trees, he also, to some self-interested extent, was a proponent of timber exploitation in Sierra Leone. The Vice President, it would seem, is an allegory of colonial forestry in Sierra Leone; he is emblematic of the different threads of forestry explored in this book.

In June 2012, in the wake of the Timbergate scandal (yet just before the WikiLeaks scandal would surface), the Vice President still persevered with his hike up the Western Area Peninsula Mountains in support of National Tree Planting Day. The ceremony was similar to previous years, with speeches by officials (government and aid donors) promoting wildlife protection and forest conversation in Sierra Leone and the symbolic planting of a tree to help promote the broader national project of 'reclothing' the Sierra Leone landscape. The Vice President and other officials were there for 'a show of their demonstration towards caring for the environment and protecting the Peninsular Forest Reserve'.[14] The Vice President again promoted the work of the German NGO Welthungerhilfe (initiated in 2009) for its European-Union funded work in enhancing conservation of the Western Area Peninsula Forest Reserve (WAPFOR). As a part of his speech, the Vice President pleaded with the local population to trust the European funders backing the re-establishment of the Forest Reserve:

> In his cautionary advice to the community people, 'the Vice President noted that if the European Union and the German people through Welthungerhilfe can come and assist in the country on any environmental project, it simply signifies that they have love for the people and government of Sierra Leone and the need for us to protect the forests for our own benefit'.[15]

The Vice President's hike can certainly be interpreted as act of environmentalism, an act of forest conservation, but ultimately one that has neocolonial shades. It is an environmentalism that does not displace a self-interested desire to exploit and profit from the country's forests, and it positions local populations as being at the source of problems with forest conservation. The National Tree Planting Day is, ultimately, a striking example of the culmination of Sierra Leone's forest conservation history.

The Vice President with his annual treks had been quite literally following in the footsteps of Sierra Leone's Forestry Department founder Charles Lane-Poole. The Western Area Peninsula Reserve was personally mapped out and overseen by Lane-Poole back in the 1910s. As John Dargavel has narrated:

> Charles Lane-Poole … on 2 November 1911 set off to demarcate his first reserve. For the next seven weeks he moved his party across the [Peninsula] mountains and up the coast, camped in the lowlands, and scrambled up hills to erect more beacons marking the boundary of the [Western Area Peninsula] reserve.[16]

The Vice President, with his re-demarcation boundary pillar, was quite literally retracing colonial lines. Like with Lane-Poole, the act of 'demarcation' was seen as being critical for forest conservation. Thus, while the aspirational objective of establishing a Forest Reserve estate that covered 30 per cent of Sierra Leone might now have faded, the modest estate of Forest Reserves that were established during the colonial era had become the key battle grounds for forest conservation in the country. Like the colonial precedent, [17] the control of Sierra Leone's forests via reserves was being justified within the moral-economic framework, as a beneficial gesture to protect the local 'ignorant' population from themselves. The legacy of *reservationism* in Sierra Leone still has resonance.

The Vice President, however, had not just been helping to re-demarcate a faded Forest Reserve; he had also been planting trees as well – often yemane (*Gmelina Arborea*) saplings no less, whose arboreal ancestors almost certainly were a part of a seed packet voyage some ninety years earlier from India to Sierra Leone. It is a tree that carries the baggage of Sierra Leone's colonial plantation history. Yemane's journey from its initial role as nurse species to its mid twentieth century role as plantation superstar – and, most recently, its value for responding to environmental reafforestation imperatives – has in different ways shaped its importance among senior politicians. With his acts of tree planting, the Vice President was reaffirming the celebration of 'National Tree Planting Day' and an approach to environmentalism that is producing a simplified forest approach – that is, a reafforestation approach that is arguably based more on enumeration than deeper philosophies of forest ecosystem health. The Vice President was not just planting trees but rather building on a *plantation* philosophy that had emerged some eighty years earlier.

In his 2012 hike, the Vice President listed the causes of deforestation. Specifically, he highlighted that 'activities such as mining, wood provision, charcoal burning, settlement and others have all contributed to the destruction of the forests and the environment in Sierra Leone'.[18] Implied in this listing, as with the forebears of colonial forestry, is that the uncontrolled and unregulated activities of 'local' populations was the most imminent threat. Large-scale logging operations were omitted from this list – that is, the particular form of forest exploitation that the Vice President had attempted to profit from in the past. There is an echo of colonial forestry logic to this scenario, an implication that *exploitation* could be a conservation objective, provided it followed a European forestry concession model. Indeed, in helping to set up a large-scale company that would export logs for profit, the Vice President's actions aligned with the longer-term objective of the colonial forestry department. It always wanted large-scale forestry operations while controlling and limiting more local-scale informal forest exploitation efforts. The Vice President's partial role in setting up Taakor Tropical Hardwoods, although an unsuccessful and potentially corrupt venture in the end, is emblematic of the *raison d'être* of colonial forestry.

The year 2012 was also a defining moment for the Western Area Forest Reserve; it was the year that it was re-gazetted and formally became the Western Area National Park. The tree space was formally redefined to emphasise the protection of its intrinsic value, and increasing restrictions on human entrance to the area were legislated, which was indicative of the changing threads of environmentalism in Sierra Leone and, more broadly, in Africa, from a resource conservation ethic to wildlife conservation. Nevertheless, this environmental shift was built on a forestry past; Forest Reserves were being rewritten as National Parks, still in the name of centralised control of forest spaces as being critical to realising conservation objectives. Thus, while the Vice President's trek might be interpreted as an environmental act focused on Sierra Leone's forest future, or on its forest present, with the imperative of improving the country's forest conservation through planting trees and re-establishing protected areas, the baggage of colonial forestry history weighed heavy, shaping the pantomime of the day's proceedings.

As this book introduced, colonial forestry history in Africa has been an increasingly rich area of scholarship, and competing theses of its legacy have emerged: on the one hand, colonial foresters have been framed as early environmentalist, laying down the institutions, laws and philosophies that would form a critical basis for the environmental movement;[19] on the other hand, they have been framed as a destructive force – misunderstanding and misreading local landscapes, dismissing existing knowledge systems – viewing them as ignorant – and ultimately creating policies that had negative (and sometimes devastating) impacts on local populations and environments.[20] Despite their seemingly contrarian positions, both of these theses have resonance and can ultimately be reconciled through questions of what environmentalism is and who it is for. Colonial forestry undoubtedly built on a sustainable resource-use ethic, protecting forests to preserve timber and to a lesser extent environmental services (e.g. protecting watersheds, preventing erosion). Environmental protection was to serve society and, more specifically, colonial society. Forest Reserves and plantations were aimed at 'protecting' and 'improving' the environment, and they certainly still form an important part of Sierra Leone's contemporary forest conservation movement. The wildlife conservation movement, while marking a shift from a resource ethic to an agenda of protecting more intrinsic forest values, has built on the early colonial forest rhetoric of deforestation blame.

Yet this is just one way to frame the story. In terms of 'environmentalism', it is an environmentalism that has been imported from Europe, and thus has been crafted from a relatively elite vision. And in this process, local populations have been largely constructed as the problem – that is, as being inherently destructive. For example, few could argue that Charles Lane-Poole, the early 'zealous conservator', did not have a passion for trying to protect forests. However, many would argue that his attitude towards the local population and their ability to manage forests was laced with deeply problematic racist overtures. Lane-Poole, among other colonial foresters, brought an assumption that European knowledge and praxis was needed to protect Sierra Leone's forests; and ultimately this meant a change in power in terms of who should control the forests. And indeed, this lies at the crux of forest conservation in Sierra Leone till this day, with the struggle to reconcile local opposition to broader environmental visions.

Environmentalism is thus contested in Sierra Leone. Historically, environmentalism was a conservationist ethic linked to the emergence of forestry. It was about wise use of forest resources based on colonial forestry 'science', an approached that was tinged with racist assumptions and that provided a broad framework for shifting control over forest resources from local populations to a central government. While on one level the more recent emergence of wildlife conservation challenges colonial forestry, it has nevertheless built upon its structures; Forest Reserves have been converted into National Parks, and the emphasis on moving forest control away from the local population has generally continued. There is therefore a third issue with environmentalism in that the narratives, policies and programmes of forest conservation in Sierra Leone have been predominantly developed without the voices of those living in the forests of Sierra Leone – that is, those who rely on the forest for their livelihoods. In this sense, the kind of environmentalism that has been predominantly pursued by colonial and postcolonial governments has not been about justice, rather it has been about security – that the forests need to be centrally controlled in order for the forests to survive and prosper.

Political ecology, at its core, is a field with a normative interest in environmental justice, as it seeks to illuminate societal winners and losers – that is, the hidden costs and power dynamics that shape different social and environment outcomes.[21] Environmental justice recognises that social and environmental dynamics are 'fundamentally interrelated, a key consequence of which is that distributions of environmental goods and ills are profoundly linked to (frequently inequitable) political-economic processes'.[22] It is therefore not surprising that marginalised rural communities in Sierra Leone have received little of the benefits but much of the burden of colonial forestry interventions. Any subsequent attempts to 'solve' this problem have tended to be shaped by similar structural political economic forces and therefore can fall into the trap of reproducing similar patterns of disempowerment and injustice.[23] Therefore, in the context of forest conservation in Sierra Leone, a critical question that needs to be asked is: who is this environmentalism for? How can empowering environmentalist objectives that improve the lives of people living in or near forests be realised? The fate of the forest is inextricably linked to their livelihoods and futures.

Currently, however, European environmental ideas and praxis are still being applied in an African context, and the resulting outcome is a not too subtle discourse of blame. Local populations are too often constructed as the problem of forest conservation. And while new, and welcome, forms of community-based approach to forest management have emerged in Sierra Leone and Africa over the past couple of decades,[24] they are strategic adjustments to, rather than radical departures from, the early colonial foundations of forest conversation laid down a century ago. Indeed, with his hikes up the Western Area Peninsula Mountains, Vice President Sam-Sumana continued the narrative that the forests of Sierra Leone need to be saved from the local population and that European conservation ideas need to displace African praxis. It is a narrative that needs to be broken and recrafted to create a truly African environmental history of forests in Sierra Leone.

Notes

1. J.D. Kandeh, 'Rogue Incumbents, Donor Assistance and Sierra Leone's Second Post-conflict Elections of 2007', *The Journal of Modern African Studies* 46 (2008), 603–35.
2. 'The Case Against Sam-Sumana', *African Confidential* 53(17) (2012), 8–9.
3. M. Mustapha, 'The 2012 General Elections in Sierra Leone: Democratic Consolidation or Semi-authoritarian', in M. Mustapha and J.J. Bangura (eds), *Democratization and Human Security in Postwar Sierra Leone* (Palgrave MacMillan, 2016), 107–30.
4. F.M. Conteh and D. Harris, 'Swings and Roundabouts: The Vagaries of Democratic Consolidation and "Electoral Rituals" in Sierra Leone', *Critical African Studies* 6(1) (2014), 57–70.
5. See J. Allouche, 'Politics, Exit Strategy and Political Settlement in Sierra Leone: A Critical Analysis of a Laboratory Experiment (1991–2015)', *Conflict, Security and Development* 17(3) (2017), 225–46.
6. WikiLeaks cable: *09FREETOWN425*. Cable date 2009. Title: VICE PRESIDENT MAY BE BOOTED FOR ALLEGEDLY. US Embassy Freetown. Classified by Political/Economic Officer Amy LeMar.
7. 'The Case Against Sam-Sumana', 8–9.
8. J. Vorrath, 'What Drives Post-War Crime? Evidence from Illicit Economies in Liberia and Sierra Leone', *Third World Thematics* 3(1) (2018), 28–45.
9. 'The Case Against Sam-Sumana', 8–9.
10. WikiLeaks cable: *09FREETOWN425*. Cable date 2009. Title: VICE PRESIDENT MAY BE BOOTED FOR ALLEGEDLY. US Embassy Freetown. Classified by Political/Economic Officer Amy LeMar.
11. 'The Case Against Sam-Sumana', 8–9.
12. 'The Case Against Sam-Sumana', 8–9.
13. 'The Case Against Sam-Sumana', 8–9.
14. 'I.K. Dumbuya VP Sumana Plants Trees and Launches WAPFOR Funded Community Dam', *Stand Times Press*, June 11, 2012.
15. 'I.K. Dumbuya VP Sumana Plants Trees'.
16. J. Dargavel, *The Zealous Conservator: The Life of Charles Lane Poole* (University of Western Australia, 2008), 29.
17. G.H. Endfield and D.J. Nash, 'Missionaries and Morals: Climatic Discourse in Nineteenth-Century Central Southern Africa', *Annals of the Association of American Geographers* 92(4) (2002), 727–42.

18. 'I.K. Dumbuya VP Sumana Plants Trees'.
19. For example, see G.A. Barton, *Empire Forestry and the Origins of Environmentalism* (Cambridge University Press, 2002); R. Grove, *Green Imperialism: Colonial Expansion, Tropical Island Edens and the Origins of Environmentalism, 1600–1860* (Cambridge University Press, 1995).
20. For example, see J. Fairhead and M. Leach, *Reframing Deforestation: Global Analyses and Local Realities with Studies in West Africa* (Routledge, 1998); N.L. Peluso, *Rich Forests, Poor People: Resources Control and Resistance in Java* (University of California Press, 1992); R.L. Bryant, *The Political Ecology of Forestry in Burma 1824–1994* (Hurst & Company, 1997).
21. P. Robbins, *Political Ecology*, 2nd edition (Wiley-Blackwell, 2012).
22. P.G. Munro, G. Van der Horst and S. Healy, 'Energy Justice for All? Rethinking Sustainable Development Goal 7 through Struggles over Traditional Energy Practices in Sierra Leone', *Energy Policy* 105 (2007), 639.
23. D. Schlosberg, 'Reconceiving Environmental Justice: Global Movements and Political Theories', *Environmental Politics* 13 (2004), 517–40; also see P.G. Munro, 'Deforestation: Constructing Problems and Solutions on Sierra Leone's Freetown Peninsula', *Journal of Political Ecology* 16 (2009), 104–24.
24. An early example of this discussion can be seen in: P.D. Cummings, 'Community Forestry – Prospects for Sierra Leone', MSc Environmental Forestry (University of Bangor, Bangor, 1990).

Bibliography

Archival Material

Harrison Institute Archives

Sevenoaks, England

Jones, T.S. *Notes on the Monkeys of Sierra Leone.* Department of Agriculture: Njala, 1950.

Jones, T.S. The Status of Large Mammals and their Conservation in Sierra Leone. Paper presented at the Wildlife Conservation in Africa, Kew, 1993.

Teleki, G. Letter to President Siaka Stevens: 24 May, 1980: Mimeo.

Teleki, G. Letter to President Siaka Stevens: 27 April, 1980: Mimeo.

Teleki, G. *Hunting and Trapping Wildlife in Sierra Leone: Aspects of Exploitation and Exportation*. Freetown: Special Report to MAF, 1980.

Teleki, G. *Outamba-Kilimi National Park: A Provisional Plan for Management and Development.* Prepared for the government of Sierra Leone and the IUCN and WWF, 1986.

The National Archives (TNA)

Richmond, England

CO 267/636/14 A. Hodson, Communication to the Natural History Museum, London: 12 February, 1932.

SH, International Colonial Office Communication: 2 March, 1932.

A. Fiddian, International communication within the Colonial Office: 4 March, 1932.

A. Hodson, Communication to the Department of Zoology: 5 August, 1932.

CO 267/679/3 R.O. Ramage, C.J. Hodgens, F.A. Baine, J. Jardine, R. Briggs, W.D MacGregor, and A.F.R Stoddart, Minutes Of A Meeting Held In The Colonial Secretary's Office On The 6th May, 1943.

FD 283/47/32 E. MacDonald, Further Report by the Conservator of Forests on the classification and enumeration of timbers, 1993.

FO 655/1212 Charles Edward Lane-Poole Passport Photo. Place of Issue: FREETOWN, 1915.

E. MacDonald, Enclosure in Dispatch, Sierra Leone No. 497 – 21st December 1933.

National Library of Australia (NLA) Archives

Canberra, Australia

MS 3799/5 C.E. Lane-Poole, Sierra Leone Diaries, 1914–1915.
C.E. Lane-Poole, Lucy the Chimpanzee. Unpublished Manuscript, 1922.
MS3799/5/1 Under Secretary of State (1910). Letter from the Colonial Office to Charles Lane-Poole – 2 August.
MS3799/6 C. E. Lane-Poole, Sierra Leone Diaries – 19 November, 1914–1915.

Rhodes House Archives (RHA)

Oxford University, England

MSS Afr r 210(1) T.E. Edwardson, Sierra Leone Diaries, 1935–1940.
MSS Brit Emp s. 459 C. Swabey, Visit of Forestry Adviser to Sierra Leone. 1957–1962.
C. Swabey, Jottings From Forestry Adviser's Tour Diaries, 1963.

Royal Commonwealth Society (RCS) Archival Collection

Cambridge, England

RCS/Y304A L. Palfreman, Christy collection of photographs on Africa: Resthouse and communal plantation, Benin City, Southern Nigeria, 1925: Cambridge University Library: Royal Commonwealth Society Archival Collection – RCS/Y304A.

School of Orientation and Africa Studies Library (SOASL) Archives

London, England

FD 30/1918 L. Palfreman, Communication from the Conservators Office, Freetown Sierra Leone – 7 March 1918.

Sierra Leone National Archives (SLNA)

Fourah Bay College, Sierra Leone

A/5/31 D.G. Thomas, Communication to E MacDonald (CoF), 1936.
A/5/45 E. MacDonald, Communication to the Colonial Secretary, 19 June 1935.
D.G. Thomas, Reopening Forestry office and stores at Kenema – Communication to the Colonial Secretary – 19 June 1935.
A/6/34 E.P. Stebbing, A Preliminary note on Farming and Forestry in Sierra Leone, 1934.
A/12/31 Forestry Department. Question of further retrenchment 1931–1932.

A/15/32 Department of Zoology, Fauna of the Gola Forest, 1932.
A/15/37 W.M. Robertson, Local Timber Exploitation Scheme, 1937.
A/24/30 K. Burbridge, Communication to the colonial office – 25 June, 1927.
A/29/32 E. MacDonald, Taungya Plantations. Advances to Farmers Communication to Colonial Treasurer, Freetown – 1 April, 1932.
A/31/3 E. MacDonald, Letter about exploitation – Communication to the Colonial Office, 1932.
A/31/32 Colonial Forest Resources Development Department. *Report on the Witte Power Saw*, 1935.
A/31/34 E. MacDonald, Memorandum on Order in Council No. 19 of 1935.
A/33/28 A. Hodson, Confidential Communication by the Governor: Protection of Fauna and Flora of Africa, 1933.
C.E. Cookson, Communication from Acting Governor to the Secretary of State for the Colonies (Sir Philip Cunliffe-Lister): August 18, 1933.
CSP, Communication to HCS: 12 April, 1934.
HCS (JM), Communication to CSP: 18 April, 1934.
A/38/41 W.D. MacGregor, Letter to the Colonial Secretary, Freetown (11 June 1941).
W.D. MacGregor, Letter to the Colonial Secretary, Freetown (10 July 1941).
A/42/38 E. MacDonald, Notes on the Committee on land deterioration in Sierra Leone.
A/43/29 E. MacDonald, Communication to the Colonial Secretary, 10 April 1929.
E. MacDonald, Government Communication on Re-organisation of the Forestry Department: 10 April 1929.
J.A. Byrne, Government Communication on Re-organisation the Forestry Department: 29 March 1929.
J.A. Byrne, Government Communication on Re-organisation the Forestry Department: 15 April 1929.
M.A. Young, Communication to Lord Passfield (Government House) – 27 July, 1929.
A/48/36 W.M. Robertson, First Half Yearly Newsletter, 1936: SLNA – E/48/30.
W.A. Robertson, Imperial Forestry Institute – Half Yearly news letter (1936).
A/51/37 D.G. Thomas, Forest Department reafforestation scheme for Bo, 1937.
A/62/31 W.B. Stanley, W.B. Memorandum, 1932.
R.R. Granville, Replies to Questionnaire regarding Wild Life in the Northern Province of Sierra Leone, 1931.
A.H. Stocks, Questionnaire regarding Wildlife in Sierra Leone: Northern Province, 1931.
A/66/39 W.F. Chipp, Governor's conference, 1939 Appointment of forest engineers for service in Nigeria, the Gold Coast, S/L, 1939: SLNA – A/66/39.
A/69/30 L. Palfreman, Communication to the Colonial Secretary: 17 August, 1918.
District Commissioner Kabala, Communication to the Colonial Secretary: 17 August, 1918.
Colonial Secretary, Communication to the Conservator of Forests: 27 September, 1918.

Colonial Secretary, Communication to District Commissioner Kabala: 18 September, 1918.
A/71/37 W.M. Robertson, Report on Forest Conference in Nigeria in 1937.
A/112/29 Commissioner of the Northern Province, Communication to the Colonial Secretary – 5th September 1929.
A/192/29 Forest Department Handing Over Notes: November, 1935. Conservator of Forest to Assistant Conservator (Mr C.V. Wallace).
AG/45/17 Colonial Secretary. Communication of Forest Reserves and Re-afforestation – 26 May 1918.
CSO 2/8 E. MacDonald, Communication to C.V. Wallace: Forestry Department handing over notes, 1929.
D/10/7 R.A. Price, Communication to the Permanent Secretary of the Ministry of Trade and Industry, 17th February 1966.
H.S. Gobeer, Communication to MTI and MANR – 'Forest Industries Corporation', 8 March 1966.
RG 3/2/92 R.M. Palmer, Communication to the Establishment Secretary – 1 May 1960.
I.G. Bulmer, Communication to Colonial Government Established Secretary – 12 September, 1960.
Public Service Commission. Communication to the Colonial Government Establishment Secretary – 23 May 1960.
Colonial Secretary. Communication on Forestry Department Senior Post Staff position, 1960.
J.S. Sawyerr, Communication to Colonial Government Established Secretary – 18 July 1963.

Published Works

Abraham, A. 'Nyagua, The British, and the Hut Tax War'. *The International Journal of African Historical Studies* 5(1) (1972), 94–98.
———. 'Bai Bureh, the British, and the Hut Tax War'. *The International Journal of African Historical Studies* 7(1) (1974), 99–106.
———. *Mende Government and Politics Under Colonial Rule: A Historical Study of Political Change in Sierra Leone 1890–1937*. Freetown: Sierra Leone University Press, 1978.
Adams, W.M. 'Nature and the Colonial Mind', in M. Adams and M. Mulligan (eds), *Decolonizing Nature: Strategies for Conservation in a Post-colonial Era* (London: Earthscan, 2002), 15–50.
Adams, W.M. and M. Mulligan. 'Introduction', in M. Adams and M. Mulligan (eds), *Decolonizing Nature: Strategies for Conservation in a Post-colonial Era* (London: Earthscan, 2002), 1–14.
African Confidential. 'The Case Against Sam-Sumana'. *African Confidential* 53(17) (2012), 8–9.
Afzelius, A. *Adam Afzelius Sierra Leone Journal 1795–1796*. Stockholm: Almqvist and Wiksell, 1967.

Akinola, B. *Authority Stealing: How Greedy Politicians and Corporate Executives Loot the World's Most Populous Black Nation*. Bloomington: Authorhouse, 2012.

Akiwumi, F.A. 'Conflict Timber, Conflict Diamonds: Parallels in the Political Ecology of 19th and 20th Century Resource Exploitation in Sierra Leone', in K. Konadu-Agyemang (ed.), *Africa's Development in the Twenty-first Century: Pertinent Socio-economic and Development Issues* (London: Ashgate, 2006), 109–125.

———. 'Environmental and Social Change in Southwestern Sierra Leone: Timber Extraction (1832–1898) and Rutile Mining (1967–2005)'. PhD Thesis. Texas State University, 2006.

Alie, J. *A New History of Sierra Leone*. Manhattan: MacMillan, 1990.

———. 'History Alive! (Final) A Reaction to Ernest Koroma's Recent Policy Statement'. *Awareness Times*, 8 August, 2007.

Alieu, E.K. 'The Birth of the First Village Forestry Association in Sierra Leone'. *Y Coedwigwr* 40 (1989), 2–4.

———. 'People's Participation in the Development and Management of Tiwai – Sierra Leone's First Game Sanctuary'. *Wildlife and Nature* 11(3) (1995), 11–21.

———. *Forestry Outlook Studies in Africa (FOSA) – Sierra Leone*. Freetown: Ministry of Natural Resources and Tourism, 2001.

———. 'Building on Local Foundations: Enhancing Local Community Support for Conservation'. *Unasylva* 61 (2010), 22–27.

———. *Country Perspective – Sierra Leone*. Shropshire: Commonwealth Forestry Association, 2011.

Allouche, J. 'Politics, Exit Strategy and Political Settlement in Sierra Leone: A Critical Analysis of a Laboratory Experiment (1991–2015)'. *Conflict, Security and Development* 17(3) (2017), 225–46.

Amanor, K.S. 'Farmers, Forestry and Fractured Environmentalism in Ghana's Forest Zones', in R.A. Cline-Cole and C. Madge (eds), *Contesting Forestry in West Africa* (Aldershot: Ashgate, 2001), 307–21.

———. 'The Symbolism of Tree Planting and Hegemonis Environmentalism in Ghana'. Paper presented at Changing Perspectives on Forests: Ecology, People and Science/Policy Processes in West Africa and the Caribbean, University of Sussex, 2001.

Animals International. 'Chimps in Danger – Sierra Leone'. 11 (1986), 19.

Arnold, J.E., G. Köhlin and R. Persson. 'Woodfuels, Livelihoods, and Policy Interventions: Changing Perspectives'. *World Development* 34 (2006), 596–611.

Asdal, H. 'The Problematic Nature of Nature: The Post-constructivist Challenge to Environmental History'. *History and Theory* 42 (2003), 60–74.

Attenborough, D. 'Expedition to Sierra Leone'. *Zoo Life 10* (1955), 11–20.

Barton, G.A. *Empire Forestry and the Origins of Environmentalism*. Cambridge: Cambridge University Press, 2002.

Bates, J.S. 'REVIEW: Sierra Leone: Report of The Forest Department for The Year 1936'. *The Empire Forestry Journal* 16 (1937), 317.

Batterbury, S. 'Political Ecology', in N. Castree, M. Hulme and J.D. Proctor (eds), *Environmental Studies* (London: Routledge, 2018), 439–42.

Beinart, W. 'African History and Environmental History'. *African Affairs* 99 (2000), 269–302.

Bennett, B. *Plantations and Protected Areas: A Global History of Forestry Management*. Cambridge: MIT Press, 2015.

Bernstein, H. and P. Woodhouse. 'Telling Environmental Change Like It Is? Reflections on a Study in Sub-Saharan Africa'. *Journal of Agrarian Change* 1(2) (2001), 283–324.

Berry, S. 'Hegemony on a Shoestring: Indirect Rule and Access to Agricultural Land'. *Africa* 62 (1992), 327–27.

Best, R. *A History of the Sierra Leone Railway 1899–1949*. Freetown: Railway Department, 1949.

Bird, M.I. and J.A. Cali. 'A Million-Year Record of Fire in Sub-Saharan Africa'. *Nature* 394 (1988), 767–69.

Blanford, H.R. 'REVIEW: Sierra Leone: Annual Report of The Forestry Department for the Year 1938. (Mimeographed.) (Government Printer, Freetown. 1939.)'. *The Empire Forestry Journal* 19 (1940), 160–61.

——. 'Moves in the Colonial Forest Service'. *The Empire Forestry Review* 26 (1947), 179.

——. 'UK Schools of Forestry, 1949–50'. *The Empire Forestry Review* 28 (1949), 109–111.

Blood, H.R. *Reorganisation and Expansion of the Forestry Department*. Freetown: Government Printer, 1941.

Blyden, S.O. 'Dr. Sama Banya: Deaf or Fen Plabar (Quarrelsome)?' *Awareness Times*, 6 December, 2011.

Bockarie, M.A. *Summary of Trees Felled by the Gava Forest Industries*. Kenema: Forestry Division, 2007.

Braidwood, S.J. *Black Poor and White Philanthropists: London's Blacks and the Foundation of the Sierra Leone Settlement 1786–1791*. Liverpool: Liverpool University Press, 1994.

Braun, B. *The Intemperate Rainforest: Nature, Culture, and Power on Canada's West Coast*. Minneapolis: University of Minnesota Press, 2002.

Braun, B. and J. Wainwright. 'Nature, Post-structuralism, and Politics', in N. Castree and B. Braun (eds), *Social Nature: Theory, Practice and Politics* (London: Blackwell, 2001), 41–63.

Brncic, T.M., B. Amarasekaran and A. McKenna. *Sierra Leone National Chimpanzee Census September 2010*. Tacugama Chimpanzee Sanctuary, 2010.

Brooks, G.E. *Landlords and Strangers: Ecology, Society, and Trade in Western Africa, 1000–1630*. Boulder: Westview Press, 1993.

Bryant, R.L. 'The Rise and Fall of Taungya Forestry: Social Forestry in Defence of the Empire'. *Ecologist* 24(1) (1994), 21–26.

——. *The Political Ecology of Forestry in Burma 1824–1994*. Hurst & Company, 1997.

Budds, J. 'Contested H20: Science, Policy and Politics in Water Resources Management in Chile'. *Geoforum* 40 (2009), 418–30.
Bulmer, I.G. *Report on the Forest Administration of Sierra Leone 1959/1960*. Freetown: Government Printer, 1960.
Burley, J.R., Mills, A., Plumptre, R.A., Savill, P.S., Wood, P.J. and Wright, H.L. 'A History of Forestry at Oxford University'. *British Scholar* 1 (2009), 236–61.
Burns, A.C. *The Nigeria Handbook: Containing Statistical and General Information Respecting the Colony and Protectorate*. Lagos: Government Printer, 1922.
Carruthers, J. '"Police Boys" and Poachers: Africans, Wildlife Protection and National Parks, the Transvaal 1902 to 1950'. *Koedoe* 36 (1993), 11–22.
———. 'Tracking in Game Trails: Looking Afresh at the Politics of Environmental History in South Africa'. *Environmental History* 11 (2006), 804–29.
Chipp, W.F. *Report on Possibilities of Development of Timber Production in Sierra Leone*. Freetown: Forestry Department, 1939.
Ciesla, W.M. and E. Donaubauer. *Decline and Dieback of Trees and Forests: A Global Overview*. Food & Agriculture Organization of the UN (FAO), 1994.
Clarke, J.I. *Sierra Leone in Maps*. London: University of London Press, 1969.
Clarke, R. 'Sketches of the Colony of Sierra Leone and its Inhabitants'. *Transactions of the Ethnological Society of London* 2 (1863), 320–63.
Clements, F.E. *Plant Succession: An Analysis of the Development of Vegetation*. Washington DC: Carnegie Inst, 1919.
———. 'Nature and Structure of the Climax'. *Journal of Ecology* 24 (1936), 252–84.
Cline-Cole, R.A. 'Towards an Understanding of Man-Firewood Relations in Freetown (Sierra Leone)'. *Geoforum* 15(4) (1984), 583–94.
———. 'The Urban Fuel Plantation in Tropical Africa: A Case for Re-evaluation'. *Land Use Policy* 7(4) (1990), 323–36.
———. 'Dryland Forestry: Manufacturing Forests & Farming Trees in Nigeria', in M. Leach and R. Mearns (eds), *The Lie of the Land: Challenging Received Wisdom on the African Environment* (Oxford: James Currey, 1996), 122–39.
———. 'Redefining Forestry Space and Threatening Livelihoods in Colonial Northern Nigeria', in R.A. Cline-Cole and C. Madge (eds), *Contesting Forestry in West Africa* (London: Ashgate, 2001), 36–63.
Cline-Cole, R., H.A.C. Main and J.E. Nichol. 'On Fuelwood Consumption, Population Dynamics and Deforestation in Africa'. *World Development* 18(4) (1990), 513–27.
Cole, N.H.A. 'Ecology of the Montane Community at the Tingi Hills in Sierra Leone'. *Bulletin de L'IFAN* 29(3) (1967), 904–23.
———. 'Ecology of a Moist Semi-deciduous Forest on Kogia Hill in Sierra Leone'. *Bulletin de L'IFAN* 30(1) (1968), 100–13.
———. 'Review of Classification of Vegetation in Sierra Leone'. *West African Science Association Journal* 13(1) (1968), 81–92.
———. *The Vegetation of Sierra Leone*. Sierra Leone: Njala University Press, 1968.
———. 'Climate, Life Forms and Species Distribution on the Loma Montane Grassland, Sierra Leone'. *Botanical Journal of the Linnean Society* 69(3) (1974), 197–210.

———. 'Toward Solving Problems of Environmental Stress in Developing Countries'. *Environmental Management* 3(6) (1979), 479–82.
———. 'The Gola Forest in Sierra Leone: A Remnant of Tropical Rain Forest in Need of Conservation'. *Environmental Conservation* 7(1) (1980), 33–40.
———. 'Floristic Associations in the Gola Rain Forests: A Proposed Biosphere Reserve'. *Journal of Pure and Applied Sciences* 2(1) (1995), 35–45.
———. 'The Role of Biodiversity Conservation in Implementing the Tropical Forestry Action Plan in Sierra Leone'. *Journal of Pure and Applied Sciences* 4 (1995), 31–40.
Colonial Office. *The Colonial Forest Service List*. London: His Majesty's Stationary Office, 1936.
Colony of Sierra Leone. *Annual Report on the Social and Economic Progress of the People of Sierra Leone, 1937*. London: His Majesty's Stationary Office, 1938.
———. *Empire Forests and the* War *Empire Forestry Conference, 1947*. Freetown: Government Printer, 1947.
Commonwealth Forestry Review, The. 'Sierra Leone'. 50 (1971), 144–45.
Conteh, F.M. and D. Harris. 'Swings and Roundabouts: The Vagaries of Democratic Consolidation and "Electoral Rituals" in Sierra Leone'. *Critical African Studies* 6(1) (2014), 57–70.
Coulthard, N. 'The RSPB Has a New Project in Africa'. *Birds Magazine*, Winter (1990), 18.
Cowen, M. 'Early Years of the Colonial Development Corporation: British State Enterprise Overseas during Late Colonialism'. *African Affairs* 83 (1984), 63–75.
Cox, P., G. Hemmant, F.M. Oliphant, R.S. Troup, A. Hill and E.P. Stebbing. 'The Encroaching Sahara: The Threat to the West African Colonies: Discussion'. *The Geographical Journal* 85(6) (1935), 519–24.
Crowder, M. *West Africa under Colonial Rule*. London: Hutchinson, 1970.
———. 'Indirect Rule: French and British Style'. *Africa* 34(3) (1994), 197–205.
Cummings, P.D. 'Community Forestry – Prospects for Sierra Leone'. MSc Environmental Forestry, University of Bangor, Bangor, 1990.
Curling, C. *Forestry Industries Department Monthly Progress Report for the Month of February 1963*. Kenema: Forest Industries Department, 1963.
Daily Mail. 'Sierra Leone Chimps for New York'. 11 May, 1955, 6–7.
———. 'Thousands of Monkeys Die in Four Days'. 7 November, 1955, 7.
———. 'Full Scale Works in Kenema Mills'. 26 November, 1957, 2.
———. 'Forestry Industry May Be Commercialised Soon'. 5 July, 1958, 1.
———. 'Expert to Review Future of Forestry Industry'. 17 July, 1958, 6.
———. 'Boys Given an Idea For a Career'. 7 May, 1959, 5.
———. 'From Man-Made Forest'. 12 November, 1959, 1.
———. 'Forest Reserves Protect Crops and Soil'. 30 December, 1959, 1.
———. 'Furniture for Sierra Leone's Hotel is Being Made Locally'. 18 May, 1960, 3.
———. 'The Story of a Sawmill in Kenema District'. 20 April, 1960, 8.
Dargavel, J. *The Zealous Conservator: The Life of Charles Lane Poole*. University of Western Australia, 2008.

Dasilva, G.L. 'An Introduction to the Tiwai Conservation Project'. *Primate Eye, 35* (1988): 24-25.

——. 'Diet of Colobus Polykomos on Tiwai Island: Selection of Food in Relation to its Seasonal Abundance and Nutritional Quality'. *International Journal of Primatology* 15(5) (1994), 655–80.

Davidson, O.R. *Energy Use Patterns, Sierra Leone.* International Development Research Centre, 1985.

Davies, A.G. *The Gola Forest Reserves, Sierra Leone: Wildlife Conservation and Forest Management.* Gland: IUCN, 1987.

Davies, A.G. and P. Richards. *Rain Forests in Mende life: Resources and Subsistence Strategies in Rural Communities around the Gola North Forest Reserve (Sierra Leone).* London: ESCOR, UK Overseas Development Administration, 1991.

Davis, D.K. 'Desert "Wastes" of the Maghreb: Desertification Narratives in French Colonial Environmental History of North Africa'. *Cultural Geographies* 11 (2004), 359–87.

Davy, J.B. 'Review: Plants of the Gold Coast by Dr F.R. Irvine'. *The Empire Forestry Journal* 10 (1931), 157–58.

Dawkins, H.C. *Report to the Forestry Division of the Ministry of Natural Resources: On Problems of Natural Regeneration, Plantation and Research.* Oxford: Commonwealth Forestry Institute, 1965.

Day, L. *Gender and Power in Sierra Leone: Women Chiefs of the Last Two Centuries.* Basingstoke: Palgrave MacMillan, 2012.

Dobson, A. '"Environmental Sustainabilities": An Analysis and a Typology'. *Environmental Politics* 5 (1996), 401–28.

Dollman, G. 'Two Skulls of a Small Race of Elephant from the Cameroons and from Sierra Leone'. *Proceedings of the Linnaean Society of London* (1933), 11–13.

——. ''Dwarf Elephants: New Skull from the Gola Forest'. *The Times*, 27 June, 1935, 12.

Dorman, M.H. *The Review of Government Departments During 1959.* Sessional Paper No.1 of 1960. Freetown: Government Printer, 1960.

Dorward, D.C. and A.I. Payne. 'Deforestation, the Decline of the Horse, and the Spread of the Tsetse Fly and Trypanosomiasis (nagana) in Nineteenth Century Sierra Leone'. *The Journal of African History* 16 (1975), 239–56.

Douay, J.E.F. 'Yemane-Gmelina Arborea (Roxb.) A Monograph, Including a Preliminary Test for Suitability as Matchwood Carried out on Plantation-Grown Material from Sierra Leone'. MSc Thesis. Imperial Forestry Institute, Oxford University, Oxford, 1954.

Douglas, S. 'The Making of Scientific Knowledge in an Age of Slavery: Henry Smeathman, Sierra Leone and Natural History'. *Journal of Colonialism and Colonial History* 9 (2008).

Drayton, R. *Nature's Government: Science, Imperial Britain, and the 'Improvement' of the World.* London: Yale University Press, 2000.

Dumbuya, K.I. 'Sierra Leone: Environmental Forum – VP Sumana Plants Trees and Lays Pillar for Western Area Peninsular Forest Reserve'. *Concord Times*, 13 June, 2011.

———. 'VP Sumana Plants Trees and Launches WAPFOR Funded Community Dam'. *Standard Times Press*, 11 June, 2012.

Dunn, L.H. 'Tree-Holes and Mosquito Breeding in West Africa'. *Bulletin of Entomological Research* 18 (1927), 139–44.

During, S.E. *Fish Smoking Trials with Acacia Tree Species as Fuel Wood*. Freetown: GTZ, 1990.

Durrell, G. *Catch Me A Colobus*. London: Penguin, 1972.

Eagleton, C. 'Sara Sophia Banks, Adam Afzelius and a Coin from Sierra Leone', in A. Craciun and S. Schaffer (eds), *The Material Cultures of Enlightenment Arts and Sciences* (Basingstoke: Palgrave Macmillan, 2016), 203–6.

Eckersley, R. *Environmentalism and Political Theory: Toward an Ecocentric Approach*. Albany: SUNY Press, 1992.

ECOWAS. *Proceedings of the Economic Community of West African States (ECOWAS) Energy Symposium*. Energy For Survival (ECOWAS, Trans.). Freetown: ECOWAS, 1982.

Empire Forestry Journal, The. 'REVIEW: Sierra Leone: Annual Report of the Forest Department for the Year 1930'. 11 (1932), 150–51.

———. 'H M The King's Birthday Honours List'. 24 (1945), 2.

———. 'Moves in the Colonial Forest Service'. 26 (1947), 179.

———. 'Sierra Leone Forest Industry'. 37 (1958), 398.

Endfield, G.H. 'Environmental History', in N. Castree, D. Demeritt, D. Liverman and B. Rhoads (eds), *A Companion to Environmental Geography* (London: Blackwell, 2009), 233–37.

Endfield, G.H. and D.J. Nash. 'Missionaries and Morals: Climatic Discourse in Nineteenth-Century Central Southern Africa'. *Annals of the Association of American Geographers* 92(4) (2002), 727–42.

———. 'Drought, Desiccation and Discourse: Missionary Correspondence and Nineteenth-Century Climate Change in Central Southern Africa'. *The Geographical Journal* 168 (2002), 33–47.

Fairbairn, W.A. 'REVIEW: Report On The Forest Administration of Sierra Leone For The Year 1947'. *The Empire Forestry Review* 29 (1950), 82.

Fairhead, J. and M. Leach. 'False Forest History, Complicit Social Analysis: Rethinking some West African Environmental Narratives'. *World Development* 23 (1995), 1023–35.

———. *Misreading the African Landscape: Society and Ecology in a Forest-Savanna Mosaic*. Cambridge: Cambridge University Press, 1996.

———. *Reframing Deforestation: Global Analyses and Local Realities with Studies in West Africa*. London: Routledge, 1998.

———. 'Reconsidering the Extent of Deforestation in Twentieth-century West Africa'. *Unasylva* 192 (1998), 38–46.

———. 'Shaping Socio-ecological and Historical Knowledge of Deforestation in Sierra Leone, Liberia and Togo', in R.A. Cline-Cole and C. Madge (eds), *Contesting Forestry in West Africa* (London: Ashgate, 2000), 64–95.

———. 'Desiccation and Domination: Science and Struggles over Environment and Development in Colonial Guinea'. *Journal of African History* 41 (2000), 35–54.

FAO. *Sierra Leone: BEFS Country Brief.* Food and Agriculture Organization of the United Nations (FAO), 2013.

FAO Release. 'Leopards Out, Monkeys In'. *The Ecologist* 1 (1978) 32.

Feika, M.B. *Report on the Forest Administration of Sierra Leone 1963/1964 to 1973/74.* Freetown: Government Printer, 1975.

Ferme, M.C. *The Underneath of Things: Violence, History, and the Everyday in Sierra Leone.* Berkeley: University of California Press, 2001.

Fimbel, C. 'Cross-Species Handling of Colobine Infants'. *Primates* 33(4) (1992), 545–49.

———. 'Ecological Correlates of Species Success in Modified Habitats May Be Disturbance-and Site-Specific: The Primates of Tiwai Island'. *Conservation Biology* 8 (1994), 106–113.

———. 'The Relative Use of Abandoned Farm Clearings and Old Forest Habitats by Primates and a Forest Antelope at Tiwai, Sierra Leone, West Africa'. *Biological Conservation* 70(3) (1994), 277–86.

Fitter, R. and P. Scott. *The Penitent Butchers the Fauna Preservation Society, 1903–1978.* London: Collins, 1978.

Forest Industries Corporation. *Sierra Leone Forest Industries Corporation, Annual Report and Accounts for the Year Ended 31st December 1966.* London: BrownKnight and Truscott Ltd, 1967.

Forsyth, T. 'Politicising Environmental Explanations: What Can Political Ecology Learn from Sociology and Philosophy of Science?' in J. Goldman, P. Nadasdy and M.D. Turner (eds), *Knowing Nature: Conversations at the Intersection of Political Ecology and Science Studies* (Chicago: University of Chicago Press, 2011), 31–46.

Fox, J.E.D. 'The Growth of *Gmelina arborea* Roxb. (Yemane) in Sierra Leone'. *The Commonwealth Forestry Review* 46 (1967), 138–44.

———. 'Exploitation of the Gola Forest'. *Journal of West African Science Association* 13(2) (1968), 185–210.

———. 'Natural Regeneration of the Kambui Hills Forest in Eastern Sierra Leone. Part I. Ecological status of the Lophira/Heritiera Rain Forest'. *Tropical Ecology* 11 (1970), 169–85.

———. 'Natural Regeneration of the Kambui Hills Forest in Eastern Sierra Leone. Part II. Methods of Silvicultural Treatment and Analysis'. *Tropical Ecology* 12 (1971), 1–23.

———. 'Natural Regeneration of the Kambui Hills Forest in Eastern Sierra Leone. Part III. The Results of Silvicultural Treatment and Management Consideration'. *Tropical Ecology* 13 (1972), 139–64.

Fyfe, C. *A History of Sierra Leone.* Oxford University Press, 1962.

Gamble, D.P. 'Kenema – A Growing Town in Mende Country'. *The Bulletin – The Journal of the Sierra Leone Geographical Association* 7–8 (1964), 9–12.

———. 'Urbanisation in West Africa'. *The Bulletin – The Journal of the Sierra Leone Geographical Association* 7–8 (1964), 7–8.

Gibson, C.C. *Politicians and Poachers: The Political Economy of Wildlife Policy in Africa.* Cambridge: Cambridge University Press, 1999.

Good, K. 'Settler Colonialism: Economic Development and Class Formation'. *The Journal of Modern African Studies* 14(4) (1976), 597–620.
GoSL. *Sierra Leone Forestry Policy 2010*. Government of Sierra Leone, 2010.
Goswami, P.C. and M. Hoskins. *Assistant to Local Community Forestry*. Rome: FAO, 1980.
Grainger, A. and W. Konteh. 'Autonomy, Ambiguity and Symbolism in African Politics: The Development of Forest Policy in Sierra Leone'. *Land Use Policy* 24(1) (2007), 42–61.
Griffiths, T. 'The Humanities and an Environmentally Sustainable Australia'. *Australian Humanities Review* 43 (2007).
Grove, R. 'Colonial Conservation, Ecological Hegemony and Popular Resistances: Towards a Global Synthesis', in J.M. MacKenzie (ed.), *Imperialism and the Natural World* (Manchester: Manchester University Press, 1990), 15–50.
——. 'Imperialism and the Discourse of Desiccation: The Case of Global Environmental Concerns and the Role of the Royal Geographical Society, 1860–1880', in R.B.M. Bell and M. Heffernan (eds), *Geography and Imperialism, 1820–1940* (Manchester: Manchester University Press, 1995), 36–53.
——. *Green Imperialism: Colonial Expansion, Tropical Island Edens and the Origins of Environmentalism, 1600–1860*. Cambridge: Cambridge University Press, 1996.
Grove, R. and T. Falola. 'Chiefs, Boundaries, and Sacred Woodlands: Early Nationalism and the Defeat of Colonial Conservationism in the Gold Coast and Nigeria, 1870–1916'. *African Economic History* 24 (1996), 1–23.
Happel, R. 'Seed-eating by West African Cercopithecines, with Reference to the Possible Evolution of Bilophodont Molars'. *American Journal of Physical Anthropology* 75(3) (1988), 303–27.
Harding D.P. and R.S.O. Harding. 'A Preliminary Checklist of Birds in the Kilimi Area of Northwest Sierra Leone'. *Malimbus* 4(2) (1982), 64–68.
Harding, R.S.O. 'Primates in the Kilimi Area, North-Western Sierra Leone'. *Folia Primatologica* 42 (1984), 96–114.
Harley, J.B. 'Deconstructing the Map'. *Cartographica: The International Journal for Geographic Information and Geovisualization* 26 (1989), 1–20.
Harris, L. and H.D. Hazen, 'Power of Maps: (Counter)-Mapping for Conservation'. *ACME: An International e-Journal for Critical Geographies* 4 (2006), 99–130.
Harris, M. *The Press in English Society from the Seventeenth to Nineteenth Centuries*. Plainsboro: Associated University Presses, 1986.
Hartley, D. 'Forest Resource Use and Subsistence in Sierra Leone'. PhD Thesis. UCL, 1992.
Hayman, R.W. 'Mammals of Sierra Leone'. *Zoo Life* 10 (1955), 2–6.
Hays, S.P. *Explorations in Environmental History*. Pittsburgh: University of Pittsburgh Press, 1998.
Haywood, A.H.W. 'Sierra Leone: The Preservation of Wildlife'. *Journal of the Society for the Preservation of the Wild Fauna of the Empire* 19 (1932), 21–32.
Helms, J.A. *Society of American Foresters*. Bethesda: Society of American Foresters, 1988.

Heneson, N. 'Loophole May Allow Trade in African Chimps'. *New Scientist*, 20 October, 1983: 165.

Hiemstra-van der Horst, G. 'We Are Scared to Say No': Facing Foreign Timber Companies in Sierra Leone's Community Woodlands'. *The Journal of Development Studies* 47 (2011), 574–94.

Hiemstra-Van der Horst, G. and A.J. Hovorka. 'Reassessing the "Energy Ladder": Household Energy Use in Maun, Botswana'. *Energy Policy* 36(9) (2008), 3333–44.

———. 'Fuelwood: The "Other" Renewable Energy Source for Africa?' *Biomass and Bioenergy* 33(11) (2009), 1605–16.

Hiemstra-van der Horst, G., P.G. Munro and S.P.J. Batterbury. 'Les réseaux illégaux du pillage: La demande globale de bois et la (re)commercialisation des forêts d'Afrique de l'Ouest'. *Écologie & Politique* 42 (2011), 47–58.

Hill, P. *FAO/IUCN African Special Project: Interim Report on Sierra Leone*. IUCN, 1963.

———. 'Sierra Leone', in T. Riney and P. Hill (eds), *Conservation and Management of African Wildlife* (Rome: FAO, 1967), 72–97.

Hodgson, D.H. *Report on the Forest Administration of Sierra Leone for the Year 1947*. Freetown: Government Printer, 1948.

———. *Report on the Forest Administration of Sierra Leone for the Year 1948*. Freetown: Government Printer, 1949.

———. *Annual Report on the Forest Administration for the Year 1949*. Freetown: Government Printer, 1951.

———. *Annual Report of the Forest Department for the Year 1950*. Freetown: Government Printer, 1952.

Hodson, A. *Trekking The Great Thirst: Travel and Sport in the Kalahari*. London: T.F. Unwin, 1912.

———. *Where Lion Reign: An Account of Lion Hunting and Exploration in South West Abyssinia*. London: Skeffington and Son Ltd, 1929.

Illes, M., P.S. Savill and G. Koker. *Gola Forest Reserves Sierra Leone, Interim Management Plan*. Sandy: RSPB, 1993.

IMMUNO AG v. Moor-Jankowski, No. 15 A.D.2d 114 (1989). Supreme Court of the State of New York 1989.

Imoudou, D. 'The Real Story: The Re-Trial of Ex-Governor Lucky Igbinedion'. *Elendu Reports*, 23 February, 2011.

International Primate Protection League Newsletter. 'Chimpanzee Traffickers Denounced in Sierra Leone Press'. 3 (1978), 5.

———. 'Sierra Leone Wildlife Laws'. 5 (1978), 15.

———. 'Ten Chimpanzees Seized at Amsterdam Airport'. 6(1) (1979), 6.

. 'News from Sierra Leone'. 7 (1980), 14.

———. 'Sierra Leone Chimpanzees Threatened'. 10(2) (1983), 6–7.

———. 'Immuno Lawsuit Thrown Out – Twice More!' 17(1) (1990), 17–19.

———. 'Immuno Sues – 1984'. 20(3) (1993), 17.

International Zoo Yearbook. 'International List of Animal Dealers'. 2 (1961), 240–48.

ITMB (Cartographer). *Map of Sierra Leone and Freetown*. Vancouver: International Travel Maps and Books, 2009.

Jacobs, N.J. *Environment, Power, and Injustice: A South African History*. Cambridge: Cambridge University Press, 2003.

———. 'The Intimate Politics of Ornithology in Colonial Africa'. *Comparative Studies in Society and History* 48 (2006), 564–603.

Jalloh, A. 'The Fula and the Motor Transport Business in Freetown, Sierra Leone'. *African Economic History* 26 (1998), 63–81.

Jha, S. 'Radical Politics and Environmentalism against Taungya in Dooars'. *Economic & Political Weekly* 47(1) (2012), 112–18.

Johnson, R.G.M. Kandeh, A. Jalloh, G.C. Nelson and T.S. Thomas. 'Sierra Leone', in A. Jalloh, G.C. Nelson, T.S. Thomas, R. Zougmoré and H. Roy-Macauley (eds), *West African Agriculture and Climate Change: A Comprehensive Analysis* (Washington DC: IFPRI), 2013.

Jones, S. 'A Political Ecology of Wildlife Conservation in Africa'. *Review of African Political Economy* 33(109) (2006), 483–95.

Jones, T.S. 'Sierra Leone', in P. Grubb, T.S. Jones, A.G. Davies, E. Edberg, E.D. Starin and J.E. Hill (eds), *Mammals of Ghana, Sierra Leone and the Gambia* (Cornwall: Trendine Press, 1998), 27–45.

———. 'Appendix II – The Sierra Leone Monkey Drives', in P. Grubb, T.S. Jones, A.G. Davies, E. Edberg, E.D. Starin and J.E. Hill (eds), *Mammals of Ghana, Sierra Leone and the Gambia* (Cornwall: Trendine Press, 1998), 214–19.

Jones, T.S., A.G. Davies, E. Edberg, E.D. Starin and J.E. Hill (eds), *Mammals of Ghana, Sierra Leone and the Gambia* (Cornwall: The Trendine Press, 1998), 214–19.

JRPG. 'Sierra Leone: Annual Report of The Forest Department For 1934'. *The Empire Forestry Journal* 14 (1935), 293.

Jumpah, D.T. *Bumbuna Hydroelectric Environmental and Social Management and Biodiversity Conservation Project*. Freetown: Government of Sierra Leone, 2012.

Kabasawa, A. 'Current State of the Chimpanzee Pet Trade in Sierra Leone'. *African Study Monographs* 30(1) (2009), 37–54.

———. 'The Chimpanzees of West Africa: From "Man-Like Beast" to "Our Endangered Cousin"', in T. Matsuzawa, T. Humle and Y. Sugiyama (eds), *The Chimpanzees of Bossou and Nimba Primatology Monographs* (Berlin: Springer, 2011), 45–57.

Kabasawa, A., Rosa M. Garriga and B. Amarasekaran. 'Human Fatality by Escaped Pan Troglodytes in Sierra Leone'. *International Journal of Primatology* 29 (2008), 1671–85.

Kaifala, J. *Free Slaves, Freetown, and the Sierra Leonean Civil War*. African Histories and Modernities. New York: Palgrave Macmillan, 2017.

Kamara, B.M. 'Letter to District Forestry Office "Timber License to Gava Forest Industries Co-operation [sic]"'. NaCEF, 2005.

Kamara, B.N. *Biodiversity: Strategic Action Plan*. Freetown Government of Sierra Leone, 2003.

———. *CBD Sierra Leone – Third National Report*. Freetown Government of Sierra Leone, 2008.

Kamara, J. *Firewood Energy in Sierra Leone – Production, Marketing, and Household Use Patterns*. Hamburg: Verlag Weltarchiv, 1986.

Kandeh, J.D. 'Ransoming the State: Elite Origins of Subaltern Terror in Sierra Leone'. *Review of African Political Economy* 26 (1999), 349–66.

———. 'Rogue Incumbents, Donor Assistance and Sierra Leone's Second Post-Conflict Elections of 2007'. *The Journal of Modern African Studies* 46 (2008), 603–35.

Koerner, L. *Linnaeus: Nature and Nation*. Cambridge, MA: Harvard University Press, 1999.

Konteh, W. 'Forest Resource Management in Sierra Leone: A Critique of Policy Formation and Implementation'. PhD Thesis. University of Leeds, Leeds, 1997.

Koroma, A.P. 'Taungya in Sierra Leone', in L.H. MacDonald (ed.), *Agro-forestry in the African Humid Tropics* (Tokyo: The United Nations University, 1981), 67–69.

Kull, C.A. 'Deforestation, Erosion, and Fire: Degradation Myths in the Environmental History of Madagascar'. *Environment and History* (2000), 423–50.

Lamb, A.F.A. *1956 Report on the Forest Administration*. Freetown: Government Printer, 1957.

———. *Report on the Forest Administration of Sierra Leone 1957*. Freetown: Government Printer, 1959.

———. *Annual Report on the Forest Administration of Sierra Leone 1958/1959*. Freetown: Government Printer, 1960.

Lane-Poole, C.E. *Report on Forests of Sierra Leone*. London: Waterlow & Sons, 1911.

———. *Report on Forest Administration for the Year 1914*. Freetown: Government Printing Office, 1915.

———. *Report on Forest Administration for the Year 1915*. Freetown: Government Printing Office 1916.

Lansana, T.E.A. *A Study of the Firewood Species Used for Fish Processing in the Tombo Village*. Freetown: GTZ, 1986.

Lavauden, L. 'Supplement: The Equatorial Forest of Africa: Its Past, Present and Future'. *Journal of the Royal African Society* 36(143) (1937), 3–25.

Leach, G. and R. Mearns, *Beyond the Woodfuel Crisis: People, Land and Trees in Africa*. London: Earthscan, 1988.

Leach, M. 'Images of Propriety: The Reciprocal Constitution of Gender and Resource use in the Life of a Sierra Leonean Forest Villages'. PhD Thesis. University of London, 1990.

———. *Rainforest Relations: Gender and Resource Use among the Mende of Gola, Sierra Leone*. Edinburgh: Edinburgh University Press, 1994.

Leach, M. and R. Mearns. 'Environmental Change and Policy', in M. Leach and R. Mearns (eds), *The Lie of the Land: Challenging Received Wisdom on the African Environment* (James Currey, Oxford, 1996), 1–33.

Linnaeus, C. *Systema Naturae*. Stockholm: Regnum Animale, 1735.
Linnard, W. 'The History of Forests and Forestry in Wales up to The Formation of the Forestry Commission'. PhD thesis. Prifysgol Bangor University, 1979.
Lloyd, A.H. 'The Operations of Military Forestry Units in France and Great Britain, 1939–44'. *The Empire Forestry Journal* 23(1) (1944), 10–19.
Lowes, R.H.G. 'Destruction in Sierra Leone'. *Oryx* 10 (1970), 309–10.
MacDonald, E. *Sierra Leone Annual Report of the Forest Department for the Year 1929*. Freetown: Government Printer, 1930.
———. *Sierra Leone Annual Report of the Forest Department for the Year 1930*. Freetown: Government Printer, 1931.
———. *Sierra Leone Annual Report of the Forest Department for the Year 1931*. Freetown: Government Printer, 1932.
———. *Sierra Leone, Annual Report of the Forest Department for the Year 1932*. Freetown Government Printer, 1933.
———. *Sierra Leone Annual Report of the Forest Department for the Year 1934*. Freetown Government Printer, 1935.
MacGregor, W.D. 'First Impressions of Forestry in Sierra Leone'. *The Nigerian Forester* 1 (1940), 20–21.
———. 'Sierra Leone'. *News Bulletin of Empire Forest Departments for the First Half of 1940* (1940), 12–13.
———. 'Sierra Leone'. *News Bulletin of Empire Forest Departments for the Second Half of 1940* (1940), 20.
———. 'Sierra Leone'. *News Bulletin of Empire Forest Departments for the First Half of 1941* (1941), 19–20.
———. *Sierra Leone Forestry Report for the Year 1940*. Freetown: Government Printer, 1941.
———. *Sierra Leone Forestry Report for the Year 1941*. Freetown: Government Printer, 1942.
———. 'The Forest Production Programme in Sierra Leone'. *Farm and Forest* 3(3) (1942), 116–17.
———. *Sierra Leone Forestry Report for the Year 1942*. Freetown: Government Printer, 1943.
———. 'Sierra Leone – A Forest Mill'. *The Empire Forestry Journal* 22 (1943), 68.
———. *Sierra Leone Annual Report of the Forestry Department for the Year 1943* (Freetown: Government Printer, 1944).
———. *Report on the Forest Administration of Sierra Leone for the Year 1944*. Freetown: Government Printer, 1945.
———. *Report on the Forest Administration of Sierra Leone for the Year 1945*. Freetown: Government Printer, 1946.
———. *Forest Department Sierra Leone: Statement of Policy*. Freetown: Government Printer, 1946.
———. *Report on the Forest Administration of Sierra Leone for the Year 1946*. Freetown: Government Printer, 1947.
MacKenzie, J.M. 'Introduction', in J.M. MacKenzie (ed.), *Imperialism in the Natural World* (Manchester: Manchester University Press, 1990), 1–14.

Maddox, G.H. 'Africa and Environmental History'. *Environmental History* (1999), 162–67.

Magbaily Fyle, C. *Historical Dictionary of Sierra Leone*. Scarecrow Press, 2006.

———. 'Pedro da çintra did not Name Sierra Leone: An Exploration into Available Evidence'. *Concord Times*, June 22, 2015.

March, E.W. 'News of Members'. *The Commonwealth Forestry Review* 47 (1968), 174.

Marsh, G.P. *Man and Nature; or Physical Geography as Modified by Human Action*. New York: Charles Scribner, 1864.

Martin, F.J. *A Preliminary Survey of Vegetation of Sierra Leone*. Njala: Department of Agriculture, 1937.

Martinez-Alier, J. *The Environmentalism of the Poor: A Study of Ecological Conflicts and Valuation*. Edward Elgar Publishing, 2003.

Matthews, S. and K. Brand. *Africa Invaded: The Growing Danger of Invasive Alien Species*. GISP, 2004.

Maughan-Brown, H. 'Sei Bush Belt in Sierra Leone'. *Farm and Forest* 4(1) (1943), 8–9.

McCann, J.C. 'The Plow and the Forest: Narratives of Deforestation in Ethiopia, 1840–1992'. *Environmental History* 2 (1997), 138–59.

———. *Green Land, Brown Land, Black Land: an Environmental History of Africa, 1800–1990*. Portsmouth, NH: Heinemann, 1999.

McGreal, S. 'Letter to the Editor: A Project with Potential to Spread Non-A, Non-B Hepatitis in West Africa'. *Journal of Medical Primatology* 12 (1983), 280–82.

McGriffin, H. 'History of Primate Conservation in Sierra Leone'. *International Primate Protection League Newsletter* 12(3) (1985), 3–5.

———. 'A New National Park in Sierra Leone'. *International Primate Protection League Newsletter* 12 (1985), 2–3.

Menzies, N. 'Three Hundred Years of Taungya: A Sustainable System of Forestry in South China'. *Human Ecology* 16(4) (1988), 361–76.

Millington, A.C. 'Environmental Degradation, Soil Conservation and Agricultural Policies in Sierra Leone, 1895–1984', in D. Anderson and R. Grove (eds), *Conservation in Africa: People, Policies and Practice* (Cambridge: Cambridge University Press, 1987), 229–48.

———. 'Environmental Degradation, Soil Conservation and Agricultural Policies in Sierra Leone, 1895–1984'. *International Journal of Sierra Leone Studies* 1 (1988), 5–25.

Minister of Trade and Industry. *Sale of Forest Industries Furniture on Deferred Payment Basis*. Freetown: Ministry of Trade and Industry, 1964.

Monomier, M. *How to Lie with Maps*. Chicago: University of Chicago Press, 1996.

Moor-Jankowski, J. 'Editorial Note: A Project with Potential to Spread Non-A, Non-B Hepatitis in West Africa'. *Journal of Medical Primatology* 12 (1983), 280.

Mueller-Hirth, N. 'If You Don't Count, You Don't Count: Monitoring and Evaluation in South African NGOs'. *Development and Change* 43(3) (2012), 649–70.

Munro, P.G. 'Protected Area Management and Community Development in Tiwai'. *Africa News*, 27 July, 2007.
———. 'Deforestation: Constructing Problems and Solutions on Sierra Leone's Freetown Peninsula'. *The Journal of Political Ecology* 16 (2009), 104–24.
———. 'A Critical History of Forest Conservation in Sierra Leone'. PhD Thesis. University of Melbourne, 2015.
———. 'Geza Teleki and the Emergence of Sierra Leone's Wildlife Conservation Movement'. *Primate Conservation* 29 (2015), 115–22.
Munro, P.G. and G.A. Hiemstra-van der Horst. 'Conserving Exploitation?: A Political Ecology of Forestry Policy in Sierra Leone'. *The Australasian Review of African Studies* 32(1) (2011), 59–72.
———. *The Domestic Trade in Timber and Fuelwood Products in Sierra Leone: Current Dynamics and Issues*. Freetown: FAO/EU, 2012.
———. *The Governance and Trade of Wood-based Products in and around the Kambui Hills North Forest Reserve* (Freetown: PAGE), 2012.
———. 'Breaks with the Past: Conflict, Displacement, Resettlement and the Evolution of Forest Socio-ecologies in Sierra Leone', in J. Lahai and T. Lyons (eds), *African Frontiers: Insurgency, Governance and Peacebuilding in Post-colonial States* (London: Ashgate Publishing, 2015), 119–30.
———. 'Contesting African Landscapes: A Critical Reappraisal of Sierra Leone's Competing Forest Cover Histories'. *Environment and Planning D* 24 (2016), 706–24.
Munro, P.G, G. Van der Horst and S. Healy. 'Energy Justice for All? Rethinking Sustainable Development Goal 7 through Struggles Over Traditional Energy Practices in Sierra Leone'. *Energy Policy* 105 (2007), 635–41.
Munro, P.G. and M. De Lourdes Melo Zurita. 'Aprendiendo del pasado: Impactos históricos de intervenciones en la agricultura de Sierra Leona'. *Ecología Política* 38 (2009), 92–95.
———. 'The Role of Cenotes in the Social History of Mexico's Yucatan Peninsula'. *Environment and History* 17 (2011), 583–612.
Mustapha, M. 'The 2012 General Elections in Sierra Leone: Democratic Consolidation or Semi-authoritarian', in M. Mustapha and J.J. Bangura (eds), *Democratization and Human Security in Postwar Sierra Leone* (Basingstoke: Palgrave MacMillan, 2016), 107–30.
Nair, P.K.R. *An Introduction to Agroforestry*. Dordrecht: Kluwer Academic Publishers, 1993.
Nature. 'The Forestry Department of Edinburgh University'. 106 (1921), 706–7.
Neumann, R.P. 'Dukes, Earls, and Ersatz Edens: Aristocratic Nature Preservationists in Colonial Africa'. *Environment and Planning D* 14(1) (1996), 79–98.
———. 'Forest Rights, Privileges and Prohibitions: Contextualising State Forestry Policy in Colonial Tanganyika'. *Environment and History* 3 (1997), 45–68.
———. *Imposing Wilderness: Struggles over Livelihood and Nature Preservation in Africa*. Berkeley: University of California, 1998.
———. 'The Postwar Conservation Boom in British Colonial Africa'. *Environmental History* 7 (2002), 22–47.

———. 'Nature-State-Territory: Towards a Critical Theorization of Conservation enclosures', in R. Peet and M. Watts (eds), *Liberation Ecologies: Environment, Development, Social Movements* (Oxford: Routledge, 2004), 195–218.

———. *Making Political Ecology*. London: Hodder Arnold, 2005.

Newland, H.O. *Sierra Leone: Its People, Products, Secret Society*. Johan Bales and Sons, 1916.

Nicol, A.K. *Annual Report on the Forest Administration of Sierra Leone for the Year 1952*. Freetown: Government Printer, 1953.

———. *Report on the Forest Administration of Sierra Leone for the Year 1954*. Freetown: Government Printer, 1955.

———. *Report on the Forest Administration of Sierra Leone for the Year 1955*. Freetown: Government Printer, 1956.

Nyerges, A.E. 'Ethnography in the Reconstruction of African Land Use Histories: A Sierra Leone Example'. *Africa* 66 (1996), 122–44.

NZ Truth. 'Animal Comedy'. 2 January (1928), 2.

Oates, J.F. *Action Plan for African Primate Conservation: 1986–1990: IUCN/SSC Primate Specialist Group*. Nairobi: UNEP and IUCN, 1985.

———. 'Sierra Leone's New Sanctuary'. *Oryx* 22(1) (1988), 50–51.

———. 'The Diet of the Olive Colobus Monkey, Procolobus verus, in Sierra Leone'. *International Journal of Primatology* 9(5) (1988), 457–78.

———. *Myth and Reality in the Rain Forest: How Conservation Strategies are Failing in West Africa*. Berkeley: University of California Press, 1999.

Oates, J.F. and G.L. Dasilva. 'Patterns of Frugivory in Three West African Colobine Monkeys'. *International Journal of Primatology* 20(3) (1999), 327–57.

Oates, J.F., H.W. George, A.G. Davies, P.G. Waterman, S.M. Green, G.L. Dasilva and S. Mole. 'Determinants of Variation in Tropical Forest Primate Biomass: New Evidence from West Africa'. *Ecology* 71(1) (1990), 328–43.

Oates, J.F. and G.H. Whitesides. 'Association between Olive Colobus (Procolobus Verus), Diana Guenons (Cercopithecus Diana), and other Forest Monkeys in Sierra Leone'. *American Journal of Primatology* 21(2) (1990), 129–46.

O'Brian, P. *Joseph Banks: A Life*. Chicago: University of Chicago Press, 1997.

Offen, K. 'Historical Political Ecology: An Introduction'. *Historical Geography* 32 (2004), 19–42.

Office of the President. *Sierra Leone: 12 Years of Economic Achievement and Political Consolidation under the APC and Dr Siaka Stevens. 1968–1980*. Freetown: State House, 1980.

Oryx. 'Conservationists in Sierra Leone'. 14 (1977), 18.

Oxby, C. 'Changes in Shifting Cultivation in Africa'. *Unasylva* 37 (1985).

Palfreman, L. 'The Position of the Crown Colonies – Sierra Leone'. Paper presented at the British Empire Forestry Conference, London 1920 – Proceedings, Resolutions and Summary of Statements, 1921.

Palmer, P.D. *The Mocking Stones*. London: Longman, 1982.

Paterson, Z. *Paterson Zochonis Plc: A Century of Enterprise*. Manchester: The Company, 1984.

Paul, I. *Report on the Forest Administration of Sierra Leone 1960/61*. Freetown: Government Printing Department, 1961.
———. *Progress Report 1955–60 by the Forestry Department of Sierra Leone*. British Commonwealth Forestry Conference, 1962.
Pelly, R.S. 'Forest Department – Disposal of Plants'. *The Sierra Leone Royal Gazette*, May 14, 1940, 280.
———. 'Shingles in Sierra Leone'. *Farm and Forest* 8 (1947), 18–19.
———. 'Sierra Leone'. *News Bulletin of Empire Forest Departments for 1950* (1950), 60–61.
———. *Annual Report on the Forest Administration for the Year 1951*. Freetown: Government Printer, 1952.
———. *Statement by the Forest Authority, Sierra Leone Prepared for the Sixth Commonwealth Forestry Conference, 1952*. Government Printer, 1952.
———. *Annual Report on the Forest Administration for the Year 1953*. Freetown: Government Printer, 1955.
Peluso, N.L. 'The History of State Forest Management in Colonial Java'. *Forest & Conservation History* 35(2) (1991), 65–75.
———. *Rich Forests, Poor People: Resources Control and Resistance in Java*. Berkeley: University of California Press, 1992.
———. 'Whose Woods are These? Counter-mapping Forest Territories in Kalimantan, Indonesia'. *Antipode 27* (1995), 383–406.
Peluso, N.L. and P. Vandergeest. 'Genealogies of the Political Forest and Customary Rights in Indonesia, Malaysia, and Thailand'. *The Journal of Asian Studies* 60 (2001), 761–812.
Peterson, D. and J. Goodall. *Visions of Caliban: On Chimpanzees and People*. Boston: Houghton Mifflin Company, 1993.
Plumptre, R. 'Obituary – Dr. H.C. Dawkins MBE'. *Forestry* 65(4) (1992), 479–80.
Polunin, N. and M.L. Curme. *World Who is Who and Does What in Environment Conservation*. London: Earthscan, 1997.
Potter, R.V., L. Danso and P.D. Palmer. *Community Participatory Forestry for Fuelwood Production in Western Area, Sierra Leone. Report of an Evaluation Mission*. Freetown: FAO, 1990.
Pratt, M.L. *Imperial Eyes: Travel Writing and Transcultration*. London: Routledge, 1992.
Prendergast, D.K. and W.M. Adams. 'Colonial Wildlife Conservation and the Origins of the Society for the Preservation of the Wild Fauna of the Empire (1903–1914)'. *Oryx* 37(2) (2003), 251–60.
Price, R.A. *Forest Industries Monthly Progress Report for the Month of Nov, 1962*. Kenema: Forest Industries Department, 1962.
———. *Forestry Industries Department Monthly Progress Report for the Month of May 1963*. Kenema: Forest Industries Department, 1963.
———. *Forestry Industries Department Monthly Progress Report for the Month of June 1963*. Kenema: Forest Industries Department, 1963.
———. *Forestry Industries Department Monthly Progress Report for the Month of August 1963*. Kenema: Forest Industries Department, 1963.

——. *Forestry Industries Department Monthly Progress Report for the Month of September 1963*. Kenema: Forest Industries Department, 1963.

——. *Forestry Industries Department Monthly Progress Report for the Month of October 1963*. Kenema: Forest Industries Branch, 1963.

——. *Forestry Industries Department Monthly Progress Report for the Month of April 1964*. Kenema: Forestry Industries Department, 1964.

Reeves, R.R., D. Tuboku-Metzger and R.A. Kapindi. 'Distribution and Exploitation of Manatees in Sierra Leone'. *Oryx* 22 (1988), 75–84.

Reno, W. *Corruption and State Politics in Sierra Leone*. Cambridge: Cambridge University Press, 1995.

Ribot, J.C. 'A History of Fear: Imagining Deforestation in the West African Dryland Forests'. *Global Ecology and Biogeography* 8(3–4) (1999), 291–300.

——. 'Forestry and Democratic Decentralization in Sub-Saharan Africa: A Rough Review', in L.A. German, A. Karsenty and A. Tiani (eds), *Governing Africa's Forests in a Globalized World* (London: Earthscan, 2001), 29–55.

Richards, P. 'The Sierra Leone-Liberia Boundary Wilderness: Rain Forests, Diamonds and War', in A. Asiwaju and P. Nugent (eds), *African Boundaries: Constraints, Conduits and Opportunities* (New York: Pinter, 1996), 205–21.

——. *Fighting for the Rain Forest: War Youth and Resources in Sierra Leone*. Oxford: James Currey, 1996.

Riddell, J.B. 'Internal and External Forces Acting upon Disparities in Sierra Leone.' *Journal of Modern African Studies* 23(3) (1985), 389–406.

Robbins, P. *Political Ecology*. Wiley-Blackwell, 2012.

Robertson, M. 'REVIEW: Second Annual Report Of The Forestry Adviser, 1951'. *The Empire Forestry Review* 32 (1953), 374–75.

Robertson, W.M. *Sierra Leone Annual Report of the Forest Department for the Year 1936*. Freetown: Government Printer, 1937.

——. 'Sierra Leone'. *News Bulletin of Empire Forest Departments for the First Half of 1937* (1937), 25–29.

——. 'L.A. King-Church'. *The Empire Forestry Review* 37 (1958), 5.

Robinson, P.T. 'Wildlife Trends in Liberia and Sierra Leone'. *Oryx* 11 (1971), 117–21.

Rodney, W. 'African Slavery and Other Forms of Social Oppression on the Upper Guinea Coast in the Context of the Atlantic Slave-Trade'. *The Journal of African History* 7(3) (1966), 431–43.

——. 'A Reconsideration of the Mane Invasions of Sierra Leone'. *The Journal of African History* 8(2) (1967), 219–46.

Roe, E.M. 'Except-Africa: Postscript to a Special Section on Development Narratives'. *World Development* 23 (1995), 1065–69.

Said, E. *Orientalism*. New York: Pantheon, 1978.

Sanon, A., P. Martin, J. Thioulouse, C. Plenchette, R. Spichiger, M. Lepage and R. Duponnois. 'Displacement of an Herbaceous Plant Species Community by Mycorrhizal and Non-mycorrhizal Gmelina Arborea, an Exotic Tree, Grown in a Microcosm Experiment'. *Mycorrhiza* 16 (2006), 125–32.

Savill, P.S. *The Classification of the Forest of Sierra Leone, with a Field Key for the Most Common Genera and Species*. University of Bangor, 1969.
———. 'The Composition of Climatic Forest Formations in Sierra Leone'. *The Commonwealth Forestry Review* 52(1) (1973), 67–71.
Savill, P.S. and G. Koker, *Gola Forest Reserves Sierra Leone, Interim Management Plan*. Sandy: RSPB, 1993.
Sawyerr, J.S. 'The Use of Melaleuca Leucodendron in Mangrove Soil Reclamation in Sierra Leone'. Paper presented at the Fifth Meeting of CROACUS, Freetown, 1963.
———. *Report on the Wildlife of Sierra Leone*. Forestry Department, 1963.
———. *Report on the Forest Administration of Sierra Leone 1962/1963*. Freetown: Government Printer, 1963.
———. *Ten-Year Plan of Forestry Development for Sierra Leone: 1964/5–1973/1974*. Freetown: Ministry of Natural Resources, 1964.
———. *Report on the Forest Administration of Sierra Leone for the year April, 1963 to March, 1964*. Freetown: Government Printer, 1965.
———. 'National Report on Man-made Forests (Sierra Leone)'. Paper presented at the World Symposium on Man-Made Forest and Their Industrial Importance and Eucalyptus Study Tour, Australia, 1967.
Schlosberg, D. 'Reconceiving Environmental Justice: Global Movements and Political Theories'. *Environmental Politics* 13 (2004), 517–40.
Scientific American. 'PROFILE: Jan Moor-Jankowski: Whistle-Blower's Wars'. (1997), 32–33.
Sharpe, A. 'The Hinterland of Liberia'. *Geographical Journal* 55(4) (1920), 299.
Sierra Herald. 'The Government of Sierra Leone's Indebtedness to Dr Lamarca'. 9(1) (2011).
Sierra Leone Forest Department. 'Exotics – Statement Submitted by the Forest Department, Sierra Leone, West Africa', in S.L. Forest Department (ed.), *The Fourth British Empire Forestry Conference*, 1935.
Sierra Leone Royal Gazette, The. 'Governor's Order'. April 3 (1920), 222–26.
———. 'Government Notices'. 52 (1921), 895.
———. 'Government Notices'. May 2 (1925), 383.
———. 'Private Enterprise Flourishes in Sierra Leone'. 1 (1961), 40–41.
———. 'Private Enterprise Flourishes in Sierra Leone'. 4 (1963), 80–81.
———. 'Furniture from Kenema Compares Favourably Anywhere'. 5(1) (1965), 6–8.
———. 'Sierra Leone Forest Industries Corporation'. 8(2) (1968), 38–41.
———. 'Furniture to be Exported to Liberia.' 8(1) (1968), 20 and 22.
———. 'Gola Forest to be Exploited'. 13 (1973), 110.
Simpson, D. 'A Preliminary History of the Kenema Area'. *Sierra Leone Studies (new series)* 21 (1967), 52–62.
Sivaramakrishnan, K. *Modern Forests: Statemaking and Environmental Change in Colonial Eastern India*. Stanford: Stanford University Press, 1999.
Squire, C. *Sierra Leone's Biodiversity and the Civil War: A Case Study Prepared for the Biodiversity Support Program*. Washington DC: World Wildlife Fund, 2001.

Stanley, W.B. 'Carnivorous Apes in Sierra Leone'. *Sierra Leone Studies* 2 (1919), 3–19.

———. 'Game Preservation in Sierra Leone'. *Sierra Leone Studies* 11 (1933), 2–15.

———. 'Wild Life in Sierra Leone'. *The Field* (1933), 1044.

Stanley, W.B. and E. Hodgson. *Elephant Hunting in West Africa*. London: Geoffrey Bles, 1929.

Stebbing, E.P. 'The Encroaching Sahara: The Threat to the West African Colonies'. *Geographical Journal* 85 (1936), 506–19.

———. *The Forests of West Africa and the Sahara: A Study of Modern Conditions*. London: W & R Chambers, 1937.

———. 'The Threat of the Sahara'. *Journal of the Royal African Society* 36 (1937), 3–35.

———. 'The Man-Made Desert in Africa: Erosion and Drought'. *Journal of the Royal African Society* 37(46) (1938), 3–40.

———. 'REVIEW: Calopepla leayana (LATR.) and the Possibilities of Control. By P.F. Garthwaite. (Indian Forest Record (Entomology), Vol. V, NO. 2'. New Delhi: Government of India Press, 1939. 2S. 6d.)'. *The Empire Forestry Journal 18* (1939), 320–24.

Story, F. 'West Tropical Africa'. *The Empire Forestry Journal* 9 (1930), 119–20.

Sundufu, A.J., M.S. James, I.K. Foday and T.F. Kamara. 'Influence of Community Perceptions towards Conservation and Eco-tourism Benefits at Tiwai Island Wildlife Sanctuary, Sierra Leone'. *American Journal of Tourism Management* 1(2) (2012), 45–52.

Sumberg, J. and M. Burke, 'People, Trees and Projects: A Review of CARE's Activities in West Africa'. *Agroforestry Systems* 15(1) (1991), 65–78.

Swift, J. 'Desertification: Narratives, Winners & Losers', in M. Leach and R. Mearns (eds), *The Lie of the Land: Challenging Received Wisdom on the African Environment* (Oxford: James Curry, 1996), 73–90.

Swyngedouw, E. 'The City as a Hybrid: On Nature, Society and Cyborg Urbanization'. *Capitalism Nature Socialism* 7 (1996), 65–80.

Sydney Morning Herald. 'Famous Monkeys – Arrival From America'. 15 July, 1927.

Teleki, G. 'Report from Sierra Leone'. *Traffic (International) Bulletin* 11 (1980), 78.

Teleki, G. and L. Baldwin. 'Sierra Leone's Wildlife Legacy'. *Zoonooz* 54(10) (1981), 21–27.

Teleki, G. and I. Bangura. 'Outamba-Kilimi National Park: Cornerstone for Conservation'. *Zoonooz* 54(10) (1981), 281–83.

Teleki, G.A., G. Davies and J. Oates, 'Sierra Leone', in R. East (ed.), *Antelopes: Global Surveys and Regional Action Plans. Part 3: West and Central Africa* (Gland: IUCN, 1990), 40–46.

Thomas, D.G. 'Report: The Forest Authority, Sierra Leone'. Paper presented at the British Empire Conference, London, 1924.

———. *Sierra Leone Annual Report of the Forest Department for the Year 1935*. Freetown: Government Printer, 1936.

Thompson, H.N. 'The Forests of Southern Nigeria'. *Journal of the Royal African Society* 10 (1911), 121–45.

Thompson, H.S. 'Status of White-Necked Picathartes – Another Reason for the Conservation of the Peninsula Forest, Sierra Leone'. *Oryx* 27(3) (1993), 155–58.

Thompson, R.W. *Wild Animal Man: Being the Story of the Life of Reuben Castang.* New York: W. Morrow & Company, 1934.

Tilley, H. *Africa as a Living Laboratory: Empire, Development, and the Problem of Scientific Knowledge, 1870–1950.* Chicago: University of Chicago Press, 2011.

Tinsley. T.W. and D.K. Barrett. *Report on a Visit to Sierra Leone to Investigate the 'Die-Back' Condition of Gmelina Arborea.* University of Oxford: Commonwealth Forestry Institute, 1965.

Tomasi, P. *Starting Community Forestry in the Western Region of Sierra Leone through Extension and Cooperation with Non-Governmental Organizations.* Freetown: FAO, 1987.

Traffic Bulletin. 'Sierra Leone's Chimps Endangered by Commercial Exploitation'. 5 (1983), 48.

Trefflich, H. and E. Anthony. *Jungle For Sale.* New York: Hawthorn Books, 1967.

Tuboku-Metzger, D. 'Nature Conservation in Sierra Leone'. *African Wildlife News* 14(3) (1979), 12–16.

——. 'Forest Exploitation in Sierra Leone: A Tale of Devastation'. *The Ecologist* 13(6) (1983), 239–41.

——. 'Robert Lowes, 1914–1990'. *Oryx* 25(2) (1991), 118.

——. *In Danger!: Endangered Species of the World.* Oxford: Heinemann Educational Books, 1999.

Unwin, A.H. *Report on the Forest and Forestry Problems in Sierra Leone.* London: Waterlow and Sons Limited, 1909.

——. *West African Forests and Forestry.* London: T. Fisher Unwin Ltd., 1920.

Vaccaro, I., O. Beltran and P.A. Paquet. 'Political Ecology and Conservation Policies: Some Theoretical Genealogies'. *The Journal of Political Ecology* 20 (2013), 255–72.

Vaclav, E. 'Trees of Forest Plantations in Sierra Leone'. *Silvaecultura Tropica et Subtropica* 6 (1978), 65–69.

Van der Horst, G. and P.G. Munro. *Land Cover Assessment of the Kambui Hills North Forest Reserve and its Surrounds.* Freetown: Energy For Opportunity, 2012.

Van der Laan, H.L. *The Lebanese Traders in Sierra Leone, Change and Continuity in Africa.* London: Mouton, 1975.

Vandergeest, P. 'Mapping Nature: Territorialization of Forest Rights in Thailand'. *Society and Natural Resources* 9 (1996), 159–75.

Vandergeest, P. and N.L. Peluso. 'Territorialization and State Power in Thailand'. *Theory and Society* 24 (1995), 385–426.

——. 'Empires of Forestry: Professional Forestry and State Power in Southeast Asia, Part 1'. *Environment and History* 21 (2006), 31–64.

——. 'Empires of Forestry: Professional Forestry and State Power in Southeast Asia, Part 2'. *Environment and History* 12(3) (2006), 359–93.

Vandi, M. 'Sierra Leone: On Government's Ban on Timber Logging and Exportation'. *Concord Times*, 28 February, 2008.

Von Hellermann, P. 'Things Fall Apart? Management, Environment and Taungya Farming in Edo State, Southern Nigeria'. *Africa* 77(3) (2007), 371–92.

———. *Things Fall Apart?: The Political Ecology of Forest Governance in Southern Nigeria*. New York: Berghahn Books, 2013.

Von Hellermann, P. and U. Usuanlele. 'The Owner of the Land: The Benin Obas and Colonial Forest Reservation in the Benin Division'. *The Journal of African History* 50(2) (2009), 223–46.

Vorrath, J. 'What Drives Post-war Crime? Evidence from Illicit Economies in Liberia and Sierra Leone'. *Third World Thematics* 3(1) (2018), 28–45.

Wadsworth, R.A. and A.R. Lebbie. 'What Happened to the Forests of Sierra Leone?' *Land* 8(5) (2019), 80.

Walker, P.A. and P.E. Peters. 'Maps, Metaphors, and Meanings: Boundary Struggles and Village Forest use on Private and State Land in Malawi'. *Society & Natural Resources* 14 (2001), 411–24.

Waugh, J. *Descriptive Information for Outamba-Kilimi National Park*. Republic of Sierra Leone, 1986.

Weiner, D. 'A Death-Defying Attempt to Articulate a Coherent Definition of Environmental History'. *Environmental History* 10 (2005), 404–20.

Weir, A.H. 'Sierra Leone: Report on Forest Administration for the Year 1939'. *The Empire Forestry Journal* 19 (1940), 331–32.

Whatmore, S. *Hybrid Geographies: Natures Cultures Spaces*. London: Sage, 2002.

Whitesides, G.H. 'Nut Cracking by Wild Chimpanzees in Sierra Leone, West Africa'. *Primates* 26(1) (1985), 91–94.

———. 'Interspecific Associations of Diana Monkeys, Cercopithecus Diana, in Sierra Leone, West Africa: Biological Significance or Chance?' *Animal Behaviour* 37 (1989), 760–76.

Whitesides, G.H., J.F. Oates, S.M. Green and R.P. Kluberdanz. 'Estimating Primate Densities from Transects in a West African Rain Forest: A Comparison of Techniques'. *The Journal of Animal Ecology* 57(2) (1988), 345–67.

Wicker, E.R. 'The Colonial Development Corporation (1948–54)'. *The Review of Economic Studies* 23(3) (1955), 213–28.

Wilkinson, A.F. 'Areas to Preserve in Sierra Leone'. *Oryx* 12(5) (1974), 596–98.

Wills, G.E.J. 'After Elephants in the Sierra Leone Protectorate and French Guinea'. *Baily's Magazine* 619 (1911), 184–91.

Wilson, J.L. *Western Africa: Its History, Condition and Prospects*. Sampsons Low, 1856.

WCMC. *Protected Areas of the World: A Review of National Systems. Volume 3: Afrotropical*. Gland: IUCN, 1991.

Wood, D. *The Power of Maps*. New York: Guilford Press, 1992.

Wood, D. and L. Fels. *Rethinking the Power of Maps*. New York: Guilford Press, 2010.

Worboys, M. 'The Imperial Institute: The State and the Development of the Natural Resources of the Colonial Empire, 1887–1923', in J.M. MacKenzie (ed.), *Imperialism in the Natural World* (Manchester: Manchester University Press, 1990), 164–86.

Zug, G.T. 'Amphibians and Reptiles of the Outamba-Kilimi Region, Sierra Leone'. *The Journal of the Herpetological Association of Africa* 33 (1987), 1–4.

Index

www.ingramcontent.com/pod-product-compliance
Ingram Content Group UK Ltd.
Pitfield, Milton Keynes, MK11 3LW, UK
UKHW021009120926
8556IPUK00029B/424